DR SHARON MOALEM

Sharon Moalem is an award-winning physician, scientist and *New York Times* bestselling author. His work has brought together evolution, genetics, biology, and medicine to explain how the human body works. His nonfiction works include the *New York Times* bestseller *Survival of the Sickest* and *How Sex Works*.

Inheritance

DR SHARON MOALEM

SCEPTRE

First published in Great Britain in 2014 by Sceptre

An imprint of Hodder & Stoughton
An Hachette UK company

First published in paperback in 2015

1

A CIP catalogue record for this title is available from the British Library

Paperback ISBN 978 1 444 76323 2
eBook ISBN 978 1 444 76324 9

Printed and bound by Clays Ltd, St Ives plc

Hodder & Stoughton policy is to use papers that are natural, renewable
and recyclable products and made from wood grown in sustainable
forests. The logging and manufacturing processes are expected to
conform to the environmental regulations of the country of origin.

Hodder & Stoughton Ltd
338 Euston Road
London NW1 3BH

www.sceptre.co.uk

For Shira

CONTENTS

CONTENTS

INTRODUCTION

Everything Is About to Change

Remember the seventh grade?

Can you recall the faces of your fellow students? Can you summon the names of the teachers, the secretary, and the principal? Can you hear the way the bell sounded? How about the smell of the cafeteria on sloppy joe day? The ache of your first crush? The panic of finding yourself in the bathroom at the same time as the school bully?

Maybe it's all strikingly clear. Or maybe, over time, your middle school years have been lost in the fog of so many other childhood memories.

Either way, you're carrying it all with you.

For a long time now, we've understood that we shoulder our experiences in the knapsack of our psyche. Even things you cannot consciously recall are somewhere in there, swimming around in your subliminal mind, ready to emerge unexpectedly for good or ill.

But it's all much deeper than that, because your body is in a constant state of transformation and regeneration and your experiences, no matter how seemingly inconsequential, from bullies to crushes to sloppy joes, have all left an indelible mark within you.

And more importantly, within your genome.

Of course, this isn't how most of us have been taught to think about the three-billion-letter equation that makes up our genetic *inheritance*. Ever since Gregor Mendel's mid-nineteenth-century[*] investigations into the inherited traits of pea plants were used to set the foundations for our understanding of genetics, we've been taught that who we are is a resolutely predictable matter of the genes we've inherited from previous generations. A little from Mom. A little from Dad. Whip it up, and there's you.

That calcified view of genetic inheritance is what students in middle school classrooms are still studying to this day when they map out pedigree charts in an effort to make sense of their fellow students' eye color, curly hair, tongue rolling, or hairy fingers. And the lesson, delivered as though on stone tablets from Mendel himself, is that we don't have much of a choice in the matter of what we get or what we give, because our genetic legacy was completely fixed when our parents conceived us.

But that's all wrong.

Because right now, whether you are seated at your desk sipping a coffee, slumped into a recliner at home, riding a stationary bike at the gym, or orbiting the planet on the International Space Station, your DNA is being constantly modified. Like thousands upon thousands of little light switches, some are turning on while others are turning off, all in response to what you're doing, what you're seeing, and what you're feeling.

This process is mediated and orchestrated by how you live, where you live, the stresses you face, and the things you consume.

---

[*] Gregor Mendel presented his work to the Brünn Natural History Society on February 8 and March 8, 1865. He went on to publish his results a year later in the *Proceedings of the Natural History Society of Brünn*. His paper was only translated into English in 1901.

And all of those things can be changed. Which, in very certain terms, means you can change. Genetically.*

This is not to say that our lives are not also shaped by our genes. They most certainly are. In fact, what we're learning is that our genetic inheritance—every last nucleotide "letter" that makes up our genome—is instrumental and influential in ways that even the most fanciful science-fiction writer could not have imagined just a few short years ago.

Day by day, we're gaining the tools and knowledge we need to embark on a new genetic journey—to take hold of a timeworn chart, lay it out across the table of our lives, and mark upon it a new course for ourselves, our children, and everyone down the line. Discovery by discovery, we're coming to better understand the relationship between what our genes do to us and what we do to our genes. And this idea—this *flexible inheritance*—is changing everything.

Food and exercise. Psychology and relationships. Medication. Litigation. Education. Our laws. Our rights. Long-held dogmas and deeply felt beliefs.

Everything.

Even death itself. Until now, most of us have been under the assumption that our life experiences end when our lives end. That's wrong, too. We are the culmination of our life experience as well as the life experiences of our parents and ancestors. Because our genes don't easily forget.

War, peace, feast, famine, diaspora, disease—if our ancestors went through it and survived, we've inherited it. And once we've got it, we're that much more likely to pass it on to the next generation in one way or another.

---

* This can include everything from acquired mutations and even small epigenetic modifications that can change the expression and repression of your genes.

That might mean cancer. It might mean Alzheimer's disease. It might mean obesity. But it might also mean longevity. It might mean grace under fire. And it might just mean happiness itself.

For better or for worse, we are now learning that it is possible to accept and reject our inheritance.

This is a guidebook for that journey.

In this book, I'm going to talk about the tools I use as a physician and scientist to apply the latest advances in the field of human genetics to my daily practice. I'll introduce you to some of my patients. I'll dig through the clinical landscape for examples of research that is important to our lives, and I'll tell you about some of the research I'm involved in. I'll talk about history. I'll talk about art. I'll talk about superheroes, sports stars, and sex workers. And I'll make connections that will change the way you look at the world and even the way you look at yourself.

I'll encourage you to walk along the tightrope that demarcates the border between the known and unknown. Sure, it's wobbly up there, but it's worth it. For one thing, the view is unforgettable.

Yes, the way I see the world is unconventional. By using genetic diseases as a template to understand our basic biology, I've made groundbreaking discoveries in seemingly unrelated fields. This approach has served me well and has led to my discovery of a new and novel antibiotic named Siderocillin that specifically targets superbug infections as well as to the granting of twenty patents worldwide for new biotechnological innovations aimed at improving our health.

I have also had the good fortune to collaborate with some of the best doctors and researchers on the planet, and I've been privy to some of the rarest and most complex genetic cases anyone has ever seen. Over the years, my career has brought me into the lives of hundreds of people who have entrusted me with the most important thing in their world—their children.

In short: I take this stuff seriously.

That doesn't mean this is going to be a grim experience. Yes, some of the things we're going to talk about will be heartbreaking. Some of these concepts may challenge many of our core beliefs. Still other ideas may be downright frightening.

But if you open yourself up to this amazing new world, it can reorient you. It might make you think about the way you live. It might just make you reconsider how, genetically speaking, you got to this very moment in your life.

I assure you: By the end of this book, your entire genome and the life it has helped shape for you will never look or feel the same again.

So, if you're ready to see genetics in a very different way, I'd like to be your guide on this journey through diverse places in our shared past, across a confounding collection of moments in our present, and into a future rife with promise and pitfalls.

In doing so, I'm going to invite you into my world and show you how I view our genetic inheritance. To start, I'm going to tell you how I think, because once you know how geneticists think you'll be better prepared for the world we're rocketing into.

And let me tell you—it is one immensely exciting place. You've opened this book at the onset of a tremendous time of discovery. Where did we come from? Where are we going? What did we get? What will we give? All of these questions are up for grabs.

This is our immediate and inexorable future.

This is our *Inheritance*.

CHAPTER 1

How Geneticists Think

For a while there, it seemed like all of New York's restaurateurs were chasing their customers' diets into a rabbit hole of vegetarian, gluten-free, thrice-certified-organic labyrinth of healthiness. Menus came with asterisks and footnotes. Servers became experts in appellations of origins, flavor pairings, and fair trade certifications as well as a muddled medley of various fats and all those confounding omegas that are good for this and bad for that.

But Jeff[1] didn't budge. Well-trained and perfectly aware of the ever-changing palates of his city's restaurant class, the young chef wasn't against healthy eating—he just didn't figure that good-for-you menus were supposed to be his top priority. So while everyone else was experimenting with freekeh and chia seeds, Jeff was cooking up big, mouthwatering, and enchantingly delicious helpings of meat, potatoes, cheese, and a whole bunch of other artery-clogging morsels seemingly made in heaven.

Your mother probably told you to practice what you preach. Jeff's mom always told him to eat what he cooked. And he did. Boy, did he ever.

But when his blood work began to show signs of higher levels of low-density lipoprotein cholesterol—the type associated with an increased risk of heart disease, often simply known as LDL—it was time to make a change. When Jeff's doctor learned the young chef also had a significant family history of cardiovascular disease, he was adamant that change should happen fast. Without a substantial modification to Jeff's diet, including a hefty daily increase of fruits and vegetables, the doctor reasoned, the only recourse to reduce his risk of a future heart attack would be medication.

That wasn't a hard verdict for the doctor to render—it was the same guidance he'd been trained to give every patient he'd ever seen with Jeff's family background and LDL presentation.

Jeff resisted at first. After all, having been given the nickname "The Steak" by others in the restaurant industry to describe his prodigious cooking and eating habits, switching to more fruits and vegetables would, he thought, hurt his reputation. Eventually, prompted by a beautiful young fiancée who wanted to grow old with him, he relented. Using his culinary training and flair for reductions, he decided to commence this new chapter of his life by first introducing fruits and vegetables into his daily repertoire, which necessitated hiding certain ones he didn't particularly enjoy eating on their own. Like parents on a health kick who conceal zucchini in their kids' breakfast muffins, Jeff started using a lot more fruits and vegetables in his glazes and reductions to go along with his black and blue porterhouse steaks. Soon enough, more than just theoretically understanding the idea of dietary balance that his doctor had preached, Jeff was living it. Smaller portions of red meat. Much bigger helpings of fruits and veggies. Sensible breakfasts and lunches.

After three long years of "eating right," and with ever-lowering cholesterol levels, Jeff figured he'd beaten back his medical problems. He was proud of himself for getting his health under control through diet—which for most people is no small accomplishment.

After strictly sticking to his new diet he figured he should feel great, but the truth was, he felt worse. Instead of increased vitality, he started to feel bloated, nauseated, and tired. An investigation into his symptoms first revealed some mild liver function test abnormalities, which quickly proceeded to an abdominal ultrasound, then to an MRI, and eventually a liver biopsy—which revealed cancer.

That was a surprise to everyone—especially his doctor—because Jeff wasn't infected with hepatitis B or C (which can cause liver cancer). He wasn't an alcoholic. He hadn't been exposed to any toxic chemicals. He hadn't done anything typically associated with liver cancer in such a young and relatively healthy person. All he'd done was change his diet, just like the doctor ordered. Jeff couldn't believe what was happening.

FOR MOST people, fructose is what gives fruit that extra sweet zing. But if you, like Jeff, suffer from a rare genetic condition called *hereditary fructose intolerance*, or HFI, you cannot fully break down fructose from your diet.[*] This causes a buildup of toxic metabolites within the body—especially within the liver—because you can't produce enough of an enzyme called fructose-bisphosphate aldolase B. And that means that for people like Jeff, an apple a day isn't healthy, it's deadly.

Thankfully, Jeff's cancer was identified early and it was treatable. A change in diet—the right one this time, away from fructose—means he will be tantalizing Gotham's palates for a long time to come.

Not everyone who has HFI is so fortunate, though. Many people with this condition spend their lives complaining of the same nausea

---

[*] It's not only fructose that's a problem but sucrose and sorbitol (which are converted to fructose in the body) as well. The latter is usually found in products such as "sugar-free" chewing gum.

and bloating Jeff experienced whenever he ate a lot of fruit and vegetables, but they never really know why. Most of the time no one takes them seriously—not even their doctors.

Not until it's too late.

Several people with HFI develop a natural strong—and therefore protective—dislike for fructose at some point in their lives and learn to avoid foods containing this sugar, even though they don't know exactly why. As I explained to Jeff when we met a short time after he finally learned of his genetic condition: When people with HFI don't listen to what their body is trying to tell them—or worse, when they're given explicit medical advice to the contrary—they may eventually suffer seizures, coma, and an early death from organ failure or cancer.

But luckily, things are changing. And fast.

It wasn't so long ago that no one—not even the richest person in the world—could get a peek into their genome. The science simply wasn't there. Today, though, the cost of exome or whole genome sequencing, an invaluable genetic snapshot of the millions of nucleotide "letters" that make up our DNA, is less than the cost of a high-quality wide-screen TV.[2] And it's getting cheaper by the day. A veritable flood of never-before-seen genetic data has arrived.

What's hidden in all of those letters? Well, for starters, information that Jeff and his doctor could have used to make more accurate decisions about how to deal with HFI and his high cholesterol—information we all can use to make individualized decisions about what to eat and what to avoid. With that knowledge—a personally monogrammed gift from every relative who ever existed before you—you'll be empowered to make educated decisions about what you eat and, as we'll explore later, how you choose to live.

None of this is to suggest that Jeff's first doctor did anything wrong—at least not in the traditional way of medical thinking. You see, from the time of Hippocrates, physicians have based their diagnoses on how their previous patients looked when they were sick. In more recent years, we've expanded this concept to include sophisticated studies that help physicians understand what remedies work best for the greatest number of people, right down to painstaking statistical percentiles.

And, indeed, that's fine. For most of the people. Most of the time.[*]

But Jeff wasn't like most people. Not even some of the time. And neither are you. None of us are.

It's been more than a decade since the first human genome was sequenced. Today, people around the world have had all or part of their genome exposed in this way, and it has become clear that no one—and I mean absolutely no one—is "average." In fact, in one research project I was recently involved with, people identified as "healthy" for the purposes of creating a genetic baseline *always* had some type of variation[**] in their genetic sequence that was out of place against what we've previously considered. Often, these variations can be "medically actionable," meaning we already know what it is and have some idea of what can be done about it.

Now, not everyone's genetic variances are likely to be as profoundly impactful on their lives as Jeff's was to his. But that doesn't mean we should simply ignore those differences—particularly not now that we have the tools to see them, evaluate them, and, increasingly, intervene in very personalized ways.

Yet not every physician has the tools and training to take those steps on behalf of their patients. Through no fault of their own, many health-care practitioners, and therefore their patients, are being

---

[*] We'll discuss this concept in far greater depth in chapter 6.

[**] Because we're medically unsure of the clinical outcomes of some of these changes, we call some of these differences variants of unknown significance.

left behind as scientific discoveries change the way we think about treating illness.

Compounding the challenge we doctors face, it's no longer enough to understand genetics. Today, physicians must also contend with *epigenetics*—the study of how genetic traits can change and be changed within a single generation and even be passed down to the next as well.

An example of this is called *imprinting*, whereby it seems that which parent, your mother or father, you inherited a certain gene from can be more important than the actual gene itself. Prader-Willi and Angelman syndromes are illustrations of this type of inheritance. On the surface they appear to be completely separate conditions, which in fact they are. However, if you dig a little deeper genetically, you'll discover that depending on which parent you've inherited imprinted genes from, you can end up with one condition or the other.

In a world in which the simplistic binary laws of genetic inheritance written by Gregor Mendel in the mid-1800s have long been treated as dogma, many physicians feel unprepared as the fast-emerging world of twenty-first-century genetics whizzes by like a bullet train speeding past a horse-drawn carriage.

Medicine will catch up eventually. It always does. But until that happens (and, frankly, even after it happens), wouldn't you like to be armed with as much information as possible?

Good. That's why I'm going to do for you what I did for Jeff when I first met him. I'm going to give you an examination.

I'VE ALWAYS found the best way to learn something is by just getting in there and doing it.

So let's roll up our sleeves and get started.

No, really—I'd actually like you to roll up your sleeve. Don't worry—I'm not going to poke you with a needle to extract your blood. That's not what I'm after. My patients often think that's the first place I'll go, but they're wrong. I just want to take a good look at your arm. I'd like to feel the texture of your skin and watch you flex your elbow. And I'd like to run my fingers along your wrist and stare deep into the crevices of your palm.

With that and nothing else—no blood, no saliva, no hair sample—your first genetic examination has begun. And I already know quite a bit about you.

People sometimes figure that when physicians are interested in your genes, the first thing to examine would be your DNA. While some cytogeneticists, people who study how your genome is physically packaged, do use microscopes to take a peek at a person's DNA, that's generally only to make sure all the volumes of chromosomes in your genome are there in their entirety and in the right number and order.

Chromosomes are small—a few millionths of a meter across—but we can see them under the right circumstances. It's even possible to see if you have a small part of one of your chromosomes missing, duplicated, or even inverted. But individual genes—the teeny-tiny, super-specific sequences of DNA that help to make you who you are? That's tougher. Even under extreme magnification, DNA appears as a twisted piece of string—maybe a bit like the curled ribbon on a beautifully wrapped birthday present.

There are ways to unwrap that gift and take a look at all those little bits and pieces inside. That generally involves a process that includes heating strands of DNA to make them separate, using an enzyme to make them duplicate and terminate at a certain place, and adding chemicals to make them visible. What materializes is a picture of you that has the potential to be more revealing than any photograph, X-ray, or MRI could ever be. And that's important, because processes that get us that deep into your DNA have a vital place in medicine.

That's not what I'm interested in right now, however. Because if you know what to look for—a small horizontal crease on the earlobe or a certain curve of the eyebrow—you can quickly make a medical diagnosis by connecting a physical characteristic to a specific genetic or congenital condition.

Which is why, right now, I'm just looking at you.

If you'd like to see yourself as I see you, go grab a mirror or walk over to the bathroom and take a look at your beautiful face. We all know our faces pretty well, or at least we think we do, so let's start there.

Is your face symmetrical? Are your eyes the same color? Are they deep set? Do you have thin or full lips? Is your forehead broad? Are your temples narrow? Is your nose prominent? Do you have a very small chin?

Now look closely at the space between your eyes. Can you fit an imaginary eye between your actual eyes? If you can, you may have an anatomical feature called *orbital hypertelorism*.

Remain calm. Sometimes in the process of identifying a certain condition or physical characteristic—and certainly whenever we give something an "ism"—doctors set off alarm bells in their patients. But if your eyes are a bit hyperteloric, there's no need to worry. In fact, if your eyes happen to be a bit farther apart than most, you're in pretty company. Jackie Kennedy Onassis and Michelle Pfeiffer are among the famous people whose hyperteloric eyes set them apart from the pack.

When we are looking at faces, eyes that are ever so slightly more widely spaced are one of the things we often subconsciously think of as attractive. Social psychologists have shown that both men and women tend to rate the faces of other people as more pleasing when those people's eyes are a bit farther apart.[3] In fact, modeling agencies purposefully seek out this trait when looking for new talent, and they have for decades.[4]

Why do we equate beauty with mild hypertelorism? Well, a good explanation comes in the form of a nineteenth-century Frenchman named Louis Vuitton Malletier.

YOU LIKELY know Louis Vuitton as the maker of some of the world's most expensive and beautiful handbags as well as the founder of a fashion empire that today has become one of the world's most valuable luxury brands. When young Louis first arrived in Paris, in 1837, he had far more modest ambitions. At the age of 16 he found work as a luggage packer for rich Parisian travelers while apprenticing for a local merchant who was known for making sturdy travel trunks—the well-stickered sort that you may recall seeing in a grandparent's attic.[5]

You might think that today's baggage handlers are rough with your luggage, but by historical comparison they treat your suitcases with kid gloves. In the days of ship travel, when inexpensive new suitcases couldn't be bought at any local department store, luggage had to be able to take a really good beating. Before Louis' trunks, most were not waterproof, and they had to be made with a rounded top to promote water runoff. That made them tough to stack and even less durable. One of Louis' clever innovations was to use waxed canvas instead of leather. This not only made the trunks waterproof but easily allowed for the switch to a flat-top design, which kept the clothing and goods packed inside dry—not a small feat given the shipping conditions of the time.

But Louis had a problem: How would he make sure people not familiar with the challenges and costs associated with his trunk design knew whether the luggage they were purchasing was of quality construction? While that wasn't a big issue in Paris, where word of mouth was the only marketing a good luggage maker needed, growing the business outside of *La Ville Lumière* was a markedly tougher job.

Compounding that dilemma was a challenge that has never gone away for Louis and his progeny—knockoffs. When rival luggage makers began copying his boxy designs, sans the quality, his son Georges came up with the illustrious interlocking LV logo, one of the first brand emblems to be trademarked in France.

With that, he reasoned, buyers could know at a glance whether they were getting the real deal. The logo was shorthand for quality.

But when it comes to biological quality, people aren't born with obvious logos. And so, over the course of millions of years of evolution, we've developed other, crude ways to size someone up—ways that tell us at a glance the three important things we need to know: kinship, health, and parental suitability.

BEYOND FACIAL similarities denoting blood relations—"you know, he looks so much like his dad"—we usually give very little thought to where our faces come from. Yet the story of the formation of our facial features is a fascinating tale—a complex embryological ballet—and any small developmental misstep is forever etched in our faces for all to see. Starting around the fourth week of our embryological lives, the external part of our face begins to develop from five swellings (imagine that these are like pieces of clay that will be shaped into what will become our future face) that will eventually merge, mold, fuse, and be fashioned into a continuous surface. When these areas don't fuse smoothly and attach, an open space remains that results in a cleft.

Some clefts are more serious than others. Sometimes a cleft results in nothing more than a small divot visible on the tip of the chin. (The actors Ben Affleck, Cary Grant, and Jessica Simpson are just a few of the people who have a cleft or "dimpled" chin.) This can also happen on the nose. (Think Steven Spielberg and Gérard Depardieu.) Other times, though, a cleft can leave a large gap in the skin exposing muscle, tissue, and bone and providing an entry point for infections.

Because they are so multifaceted, our faces serve as our most important biological trademark. Just like the Louis Vuitton logo, our faces speak volumes about our genes and the genetic workmanship that went into our fetal development. For this reason, our species learned to pay attention to these clues before we even knew what they really meant, because they also provided the fastest way of assessing, ranking, and relating to the people around us. Far more than just a superficial perspective, the reason we give so much importance to how our faces look is that, whether we like it or not, they can divulge our developmental and genetic history. Your face can also tell us a lot about your brain.

Facial formation can signal whether or not your brain developed under normal conditions. In the genetic game of sizing people up, millimeters matter. That could help explain why, across many cultures and generations, we've developed a special attraction to eyes that are ever so subtly farther apart than most people's. The spacing between our eyes is a feature of more than 400 genetic conditions.

Holoprosencephaly, for instance, is a condition in which the two hemispheres of the brain don't form properly. Besides being more likely to have seizures and intellectual disability, people with holoprosencephaly are also likely to have *orbital hypotelorism*—eyes that are very closely spaced. Hypotelorism has also been associated with *Fanconi anemia*, another fairly common genetic condition in people of Ashkenazi Jewish or black South African descent.[6] This condition often causes progressive bone marrow failure and an increased risk for malignancies.

Hyper- and hypotelorism are just two road signs along a developmental highway that brings together our genetic inheritance and our physical environment, but there are other indicators to look for as well.

Let's go find some of those things.

Take another look in the mirror. Are the outer corners of your eyes lower than the inner corners? Are they higher? We call the separation between the upper and lower eyelids the palpebral fissure. If the outer corners of your eyes are higher than the inner, we describe that as upslanting palpebral fissures. In many people of Asian ancestry this is completely normal and a defining feature, but in individuals of other ancestries, substantially upslanting palpebral fissures may be one of the specific signs or indications of a genetic condition such as Trisomy 21 or Down syndrome.

When the outer corners of the eyes are lower than the inner, the term for this is downslanting palpebral fissures, which again may mean nothing on its own. But it could also be an indicator of Marfan syndrome, a genetic connective tissue disorder, as it was in the case of the late actor Vincent Schiavelli, who played Fredrickson in the film *One Flew Over the Cuckoo's Nest* and Mr. Vargas in *Fast Times at Ridgemont High*. For casting agents, Schiavelli was "the man with the sad eyes." For those who know the clues, though, those eyes were a marker that pointed, along with flat feet, a small lower jaw, and several other physical signs, to a genetic condition that when left untreated can result in heart conditions and shortened lifespans.

Another, less debilitating condition in which the same principle of discovery applies is *heterochromia iridum*, an anatomical feature in which a person's irises don't match in color. It's often the result of an uneven migration of melanocytes, the cells that produce melanin.

You might immediately think of David Bowie, since much has been made of the strikingly different appearance of his eyes. If you look closely, though, you'll see that Bowie's eyes aren't different colors but rather that one of his pupils is fully dilated—the result, it turns out, of a high school fight over a girl.

Mila Kunis, Kate Bosworth, Demi Moore, and Dan Aykroyd are a few of the people who are true members of the heterochromia club. Even though you're probably familiar with some or all of those people, you might not have noticed it before, since heterochromia is often subtle.

You probably know some people with heterochromia and never even realized it. Ordinarily, we don't spend a lot of time looking deep into the eyes of our friends and acquaintances. Nonetheless, there's probably someone in your life whose eyes are burned into your psyche.

Aside from our most significant others, though, we often only remember people's eyes if they are strikingly and brilliantly blue, like a perfectly cut aquamarine gemstone—a pretty consequence of a complete failure of pigmentation cells to go where they are supposed to go during fetal development.

And if those blue eyes are accompanied by a white forelock, I think immediately of *Waardenburg syndrome*. If you have a streak of hair without pigment, heterochromatic eyes, a wide nasal bridge, and hearing problems, then chances are good that you have this condition.

There are a few different types of Waardenburg syndrome, but the most common is Type 1. This variety of Waardenburg syndrome is caused by changes in a gene called *PAX3*, which plays a critical role in the way cells migrate as they make their way out of the fetal spinal cord.

Studying the way the gene works in people with Waardenburg

syndrome may provide insights that can be used to understand other, much more common conditions as well. *PAX3* is also thought to be involved in melanomas, the deadliest type of skin cancer—an example of how the hidden inner workings of our bodies become apparent through rare genetic conditions.[7]

Now let's move onto eyelashes. While some of us take them for granted, there's actually a whole industry to make us better endowed in this department. If you're looking for fuller lashes, you could consider getting extensions or even try using an eyelash-enhancing drug by the trade name Latisse.

But before you do any of those things, I'd like you to get a good look at your eyelashes and see if you can count more than one row. If you find a few extra eyelashes or an entire row, you have a condition called *distichiasis*. You're also in pretty famous company, Elizabeth Taylor is only one example of someone who shares your condition. Interestingly, it's thought that having an extra row of eyelashes is part of a syndrome called *lymphedema-distichiasis syndrome*, LD for short, which is associated with mutations in a gene called *FOXC2*.

The lymphedema in the disease name refers to what happens when there's less than normal fluid drainage, as when you've been sitting on a long flight and your shoes no longer fit. In this condition it's especially pronounced in the legs.

Not everyone with an extra row of eyelashes is symptomatic for swelling, though, and it's not exactly clear why. You or someone you love may have an extra row of eyelashes and never noticed it until now.

You never know what you'll find when you start looking at people in this way. Which is exactly what happened to me last year while sitting at the dinner table with my wife. I always thought that it was mascara that was giving her a very full-looking set of upper eyelashes. But I was wrong. My wife has distichiasis.

Although she doesn't have any of the other associated symptoms of LD, I couldn't believe that it took me more than five years of marriage to notice. Which puts a whole new genetic spin on the idea of finding new qualities in our spouse, even after many years. I just never thought that I could actually miss an extra row of eyelashes.

This proves that our faces can be vast and unexplored genetic landscapes. You just need to know how to look.

By now you might have identified at least one feature on your face that may be connected to a genetic condition. But chances are good that you don't, in fact, have that condition. The truth is that everyone is "abnormal" in some way, so it's rare to be able to link a single physical characteristic with a correlating condition. When such characteristics are analyzed piece by piece and combined—the spacing and slant of your eyes, the shape of your nose, the number of rows of eyelashes—a tremendous amount of information can be gained about people. And it's this gestalt that can lead us to a genetic diagnosis—one we can reach without ever having to take an in-depth look into your genome. True, the confirmation of a clinical suspicion is often done through direct genetic testing, but combing through a person's entire genome without a specific target is like sifting through every grain of sand on a beach looking for a grain that is just slightly different from the others. A daunting and onerous computational task, to be sure.

So, in short, it helps to know what you're looking for.

RECENTLY, I was at a dinner party with some of my wife's friends whom I hadn't met before. And I just couldn't stop staring at the hostess.

Susan had eyes that were slightly spread apart (hyperteloric)—just enough to be noticeable. Her nose was just a bit flatter across the bridge than most people's. She had a rather distinct, wide peak to her vermillion border (doctor shoptalk to describe the shape of her upper lip). She was also a little on the short side.

And as her hair danced along her shoulders, I was fixated on the prospect of getting a glimpse of her neck. While pretending to admire a rare French poster on the wall for François Truffaut's 1959 film *The 400 Blows* and craning my neck as inconspicuously as possible, I tried to sneak a peek.

It didn't take long for my wife to notice my blatant gawking, and she pulled me aside in a quiet hallway.

"Come on! Are you looking again?" she asked. "If you don't stop staring at Susan people are going to get the wrong idea."

"I can't help it. Remember the other day with your eyelashes?" I said. "Sometimes I just can't turn it off. Seriously, though, I think Susan has Noonan syndrome."

My wife rolled her eyes, knowing full well where this was going. I would be awful company for the rest of the night, ruminating on the various diagnostic possibilities presented by our host's physical appearance.

Here's the thing: Once you learn how to look, manners easily go out the window, and it becomes almost impossible not to. You might have heard that many doctors believe they have an ethical duty to stop and render aid to those in immediate need—at the scene of an accident before paramedics arrive, for instance. What then of those physicians who have been trained to see the possibility of serious, even life-threatening conditions, where others might not see anything unusual at all?

As I continued to study Susan's features, I had a significant ethical dilemma on my hands. The host and other guests were certainly not my patients, and they surely hadn't invited me to diagnose any possible genetic or congenital conditions they might have. This was a woman I'd only just met. How would I broach the subject? Or stop myself from blurting out that her distinctive appearance—her eyes, her nose, her lips, and possibly a trademark stretch of skin connecting her neck and shoulders, called a webbed neck—made it quite likely that she had a genetic condition. Besides having implications for any future children, Noonan syndrome is also associated with potential heart disease, learning disabilities, impaired blood clotting, and other troubling symptoms.

Noonan syndrome is just one of many so-called "hidden conditions," since the associated traits aren't all that unusual. As with the extra row of eyelashes, it's not uncommon for people to be unaware that they have it until they start looking for it. It wasn't as though I could simply walk up to her and say, "Thank you for inviting us for dinner. The tempeh was delicious. By the way, did you know that you have a potentially deadly autosomal dominant disorder?"

Instead, I decided to just ask if there were wedding pictures around. I thought this might help clarify for me if she really had Noonan syndrome, which is generally inherited from an affected parent. After the second photo album and umpteenth picture of the bride with her mother, it was clear that they shared many of the same physical characteristics.

"Yup," I thought. "Noonan it is."

"Wow," I said, going in for what I hoped would be a soft broaching of the subject. "You *really* look like your mother."

"Yes, I get that often," was her first response. "Actually, your wife told me a little bit about what you do…"

At that exact moment I really wasn't sure in which direction the conversation was headed. Mercifully, Susan came to my rescue.

"My mother and I have this genetic condition, it's called Noonan syndrome, have you heard of it?"

As it turns out, Susan was well aware of her condition, though few others were. And friends at the party, who had known her much longer than I, marveled at how I'd been able to diagnose her condition based on slight physical differences they'd hardly noticed.

The truth is, though, that it doesn't take a physician to do this sort of thing. Everybody does it. You did it the last time you saw someone with Down syndrome. You might not have thought about it as your eyes surveyed the hallmark attributes—upslanting palpebral fissures, short arms and fingers (called brachydactyly), low-set ears, a flat nasal bridge—but you were conducting a rapid genetic diagnosis. Having seen enough cases of Down syndrome over your lifetime you unknowingly ran through a mental checklist of features to arrive at a medical conclusion.[8]

We can do this with thousands of conditions. The better we get at it, the tougher it is to stop doing it. It can be annoying (as understandably it sometimes is to my wife), and it might wreck a dinner party, but it's also important—because sometimes a person's appearance is the only way to determine that they have a genetic or congenital disorder. Sometimes, believe it or not, as you'll see in a moment, we just don't have any other reliable test.

GO BACK again and take a look at the area between your nose and upper lip. Those two vertical lines demarcate your philtrum, and this happens to be the site where, during early development, several pieces of tissue migrated and met, like great continental shelves crashing together to form a mountain range.

Do you remember what I said about our faces being a lot like a Louis Vuitton logo—a sign of our genetic quality and developmental history? Now, if you're having trouble seeing the lines of your philtrum and the area is somewhat smooth, and if your eyes are a bit small or spread apart, and if you've got an upturned nose, too, then your mother may have been drinking while she was pregnant with you, creating a perfect storm of exposure called *fetal alcohol spectrum disorder*, or FASD. We tend to hear these words together and cringe, because FASD is commonly thought to be a devastating cluster of disorders. It can be. But it can also be mildly expressed, sometimes with just a few physical facial clues and little else as a result. In spite of all the amazing breakthroughs in medicine and genetics that we've experienced over the past decade, there's still no definitive test for it other than the same visual inspection that you just performed on yourself.[9]

That brings us back to your hand. Now that you have an idea of how specific traits and the combinations of those traits can provide information about someone's genetic makeup, you can look at your hand the way I would. Take a look at the lines on your palms. How many major creases do you have? I have a big curved one running opposite my thumb, and then two creases running horizontally below my fingers.

Do you have a single crease running along your palm, below your fingers? This can be associated with FASD and Trisomy 21, but rest assured because around 10 percent of the population has at least one hand with one abnormality and no other indicators of genetic disease.

What about your fingers? Are your fingers excessively long? If so, you might have *arachnodactyly*,* a condition of long fingers that can be associated with Marfan syndrome and other genetic disorders.

And as long as we're looking at your fingers, do they taper toward your nails? Are your nail beds deep set? Now have a good look at your pinkies. Are they straight, or do they curve inward toward the rest of your fingers? If they do have a distinctive curve, you might have

---

* Also called spider fingers.

something called *clinodactyly*, which can be associated with more than 60 syndromes, or be isolated and completely benign.

Don't forget your thumbs. Are they wide? Do they look like your big toes? If so, that's called *brachydactyly type D*, and if you've got it, you're in an inheritance club that includes the actor Megan Fox, although you wouldn't know it based on Motorola's 2010 Super Bowl advertisement in which she starred, because the directors used a thumb double.[10] It can also be a symptom of *Hirschsprung's disease*, a condition that can affect the way your intestines work.

You may want a little privacy for this next examination. If you're reading this book at home or somewhere else where you don't feel self-conscious, slip off your shoes and socks and gently pull apart your second and third toes. If you find that there's an extra little flap of skin there, then you likely carry a variation in the long arm of your chromosome 2 that is associated with a condition called *syndactyly type 1*.[11]

We all start out, during the first stages of our development, with hands that have the appearance of baseball gloves. But as we develop we lose the webbing between our fingers as our genes help instruct the skin cells between our fingers and toes to die off.

Sometimes, though, the cells refuse to go. On our hands and feet, that's usually not the end of the world; surgery can generally fix the rare cases of syndactyly that are debilitating—and lots of people have started to get creative with extra skin between their toes, using tattoos and piercings to call hipster-like attention to a little bit of extra epidermal terrain that most people don't have.

If you have a child with this condition who isn't yet old enough for body art, you could always tell them it might just make them a better swimmer. That's the case for ducks, of course. Ducks use their webbed feet to balance and paddle along when in the water and to thrust themselves, jet-like, when they're below the surface looking for food.

How do ducks keep their feet webbed? The tissue between their digits survives thanks to the expression of a protein called Gremlin, which behaves a little like a cellular crises counselor, convincing the cells between a duck's toes not to kill themselves, as they would in most other species of birds and people alike. Without Gremlin, it seems, ducks would have feet like chickens. And that wouldn't do them much good in the water.

Now, can you bend your thumb to touch your wrist? Can you bend your pinkie back beyond 90 degrees? If so, you might have one of a very common and underdiagnosed group of conditions called *Ehlers-Danlos syndrome*. And you may need to start taking a medication called an angiotensin II receptor blocker, which is currently under clinical investigation, to keep your aorta from dissecting (or shredding). That sounds dramatic, but yes, it's true: From a simple evaluation of your hands, you can tell if you're at increased risk of cardiovascular complications.

That's how some physicians use genetics to inform their practice. Yes, sometimes we employ high-tech tools to take a look at your genetic mural. Sometimes we stay up late at night studying your genetic sequence on an online database, like a computer programmer trying to debug a complicated piece of code. But quite often we use a combination of very low-tech techniques to diagnose conditions. And sometimes it's a combination of simple, subtle clues with high-tech analysis that tells us what we most need to know about what's going on deep and small inside of you.

WHAT DOES this look like in practice? Well, before I even lay eyes on a patient, I'm usually given a referral slip from another physician. On a good day, I get a detailed letter explaining why that physician would like me to see their patient and what specific concerns they might have. Sometimes they proffer a very educated guess.

And often not.

Usually, I'm starting with short, vague terms like "developmental delay." Other times I get a message like "hirsutism or multiple pigmented patches on skin, along the lines of Blaschko." Yes, over the years computers have eliminated the challenge of deciphering physicians' notoriously bad handwriting, but we still seem to pride ourselves on using complicated and esoteric language.

Of course, it could be worse: In the past, some physicians would note in the chart or referral F.L.K., which inappropriately meant "funny-looking kid." This was medical shorthand for "I'm not quite sure what's wrong, but something just doesn't look right." For the most part, those initials have been replaced with the more scientific, accurate, and compassionate word *dysmorphic*. But that's still a vague description.

It only takes a few short words to send my mind racing. Even before I see a patient who has been described to me as dysmorphic, I begin running through all the algorithms I've internalized and start thinking about the important things I need to remember to ask the patient and their family. I consider what few clues I already have. A patient's name sometimes gives hints of ethnic background, an important factor in many genetic diseases—and since some cultures have long histories of intrafamily marriages, names can also clue me in to the possibility of the patient's parents being related.[12] An age tells me where someone might be in the development of his or her condition. And the department from which the referral comes gives me a clue as to what the most obvious or pressing symptoms of the patient's condition might be.

This, for me, is stage 1.

Stage 2 starts as soon as I enter the examination room. You might have heard that people in charge of reviewing applicants for a new job gain a tremendous amount of information about a candidate within the first few seconds of meeting. The same goes for doctors. Almost immediately, I begin to deconstruct my patient's face, much like you examined your own face in the mirror. I look at the patient's eyes, nose, philtrum, mouth, chin, and a few other landmarks and then try to rearrange them, putting them back together piece by piece. Before I ask a patient anything at all, I ask myself, how is this person different?

Dysmorphology is a relatively young field of study that uses the parts of the face, hands, feet, and rest of the body to give us clues about an individual's genetic inheritance. Disciples of this field try to identify physical clues that reveal the presence of an inherited or transmitted condition, not unlike art experts who employ knowledge and tools to determine the authenticity of a painting or sculpture.[13]

Dysmorphology is also the first tool I retrieve from my toolbox when I'm meeting new patients. But of course, that's not where it ends. Before I'm done, I'm going to want to know a lot more about you.

That makes me a little bit different from most physicians. You see, a lot of your doctors get to know parts of you. Your cardiologist gets to see your heart in all its blood-pumping glory. Your allergist might know how you fare against pollens, environmental pollutants, and other personal poisons. Orthopedists take care of your crucial bones. Podiatrists are there for your precious feet.

But as your physician with a special interest in genetics, I'm going to see a lot more of you. I'm going to have a look at every part. Every curve. Every crevice. Every bruise. And every secret.

Locked away inside the nucleus of your cells is an encyclopedia about who you are, where you've been—and a whole bunch of clues as to where you're going. And sure, some of the locks are going to be easier to pick than others, but it's all there.

You just need to know *where* and *how* to look.

CHAPTER 2

When Genes Misbehave

What Apple, Costco, and a Danish Sperm Donor Teach Us about Genetic Expression

In the modern world of classic genetics, Ralph is Mendel's pea.

For several years, the prodigious Danish sperm donor was a sought-after provider of the basic genetic elements that would, when paired with the genetic material of eager mothers across the globe, produce a rather predictable number of tall, strapping, and fair-haired children.

And for a while, it seemed, everyone wanted a piece of that action.

At 500 Danish kroner per sample (about $85 USD), a lot of young men with the right stuff (generally a combination of desirable physical and intellectual characteristics coupled with a high sperm count) have turned to semen donation to help make ends meet in Denmark, where tolerant social attitudes and Viking allure have made human semen a popular export.[1]

But even by Scandinavian standards, Ralph was downright prolific.

Owing to concerns that unwitting siblings might accidentally meet up—and hook up—somewhere down the road, donors like Ralph were supposed to stop providing semen after siring 25 children. But no one seemed to have figured out how to know when someone's limit had been reached. And Ralph—whose dossier photo featured him riding a three-wheeled bicycle wearing Adidas shorts and a red vest—was so popular that, when he stopped donating of his own accord, some prospective parents, fixedly desirous of his genes, took to Internet message boards seeking to procure extra vials of his frozen semen.

Ultimately, the man known to most of his recipients only as Donor 7042 would become the biological father of at least 43 children in several nations.

As it turned out, though, Ralph wasn't just sowing his Nordic oats. He was unknowingly spreading a bad seed—passing along a gene that causes excess body tissue to develop with sometimes disconcerting and life-altering results, including enormous sacks of sagging skin, profound facial deformities, and growths that can resemble deep-red, body-covering boils. The tumor-producing disorder, called *neurofibromatosis type 1*, or NF1, can also cause learning difficulties, blindness, and epilepsy.

The story of Donor 7042 and his unfortunate offspring captivated public attention and resulted in swift changes to Danish laws governing the number of children who can be fathered by sperm donors.[2] But for some families it was too little, too late.

DNA had been passed. Babies had been made. Genes had been inherited. The principles first established by Gregor Mendel, the father of modern genetics, back in the mid-1800s, were alive and not so well in the twenty-first century.

So why then were Ralph's offspring afflicted with a disease from which he didn't seem to suffer?

GREGOR MENDEL wasn't all that interested in peas. Not at first, at least. Instead, the inquisitive young monk wanted to experiment on mice.

It took a dour old man named Anton Ernst Schaffgotsch to change Mendel's direction—and in doing so, Schaffgotsch changed history.

You see, if you were a monk with your eyes set on artistic endeavors or scientific discovery back in Mendel's day, you could do no better than a calling to the humble hillside monastery of St. Thomas in the city of Brünn, in what is now the Czech Republic.

The monks of St. Tom's had long been a roguish bunch of reverends. Sure, they were always mindful that their primary responsibilities lay in service to their Lord, but within the confines of the abbey's crumbling brick walls they'd developed a collegial culture of inquiry. Alongside prayer there was philosophy. Alongside meditation there was mathematics. There was music, art, and poetry.

And, of course, there was science.

Even today, their collective discoveries, insightful visions, and raucous debates would give church leaders a good case of heartburn. During the long, authoritarian reign of Pope Pius IX, however, their collective exploits were downright subversive. And Bishop Schaffgotsch was not amused.

In fact, he had only tolerated the abbey's extracurricular activity, Mendel's journals indicate, because he didn't understand much of it.

Initially, Mendel's work on the mating habits of mice seemed simple enough. But eventually, to Schaffgotsch, it simply went too far.[3] For starters, the caged rodents in Mendel's spacious, stone-floored quarters gave off a stench that Schaffgotsch found incompatible with the tidy life expected of a monk of the Augustinian order.

Then there was the sex.

Mendel, who like all of the monks at St. Thomas had taken a vow of consecrated chastity, seemed obsessively interested in how the furry little creatures were getting it on.

That, Schaffgotsch figured, was beyond the pale.

So the dour bishop ordered the inquisitive young monk to shut down his little mouse brothel. If Mendel were, as he professed, purely interested in how traits move from one generation of living creatures to the next, he'd have to be content with something less titillating.

Something like peas.

Mendel was duly amused. What the bishop didn't seem to understand, the impish monk mused, was that "plants also have sex."

And so, over the next eight years, Mendel grew and studied nearly 30,000 pea plants and discovered, through duteous observation and record keeping, that certain traits of the plants—stem size and pod color, for instance—followed particular patterns from one generation to the next. Those findings set the stage for our understanding that genes dance in pairs, and when one gene is dominant over another (or when two recessive genes come together to tango) it can prompt a specific trait.

It's impossible to say what might have happened had Mendel continued to work with mice. In studying the far more behaviorally complex creatures, he might have missed altogether the discoveries he made while seeking to better understand how to breed consistently smooth, green, and long-stemmed peas. Then again, the meticulous monk, if given more time to watch his mice mix whiskers, might very well have stumbled onto something even more revolutionary—something that took more than a century for his disciples to begin to recognize. As it happened, though, when Mendel first published his findings in an obscure journal called *Proceedings of the Natural History Society of Brünn*, his work was met with a collective scientific "meh." And by the time it was rediscovered at the turn of the twentieth century, he'd long since been buried in the city's Central Cemetery.

But like many visionaries whose work is not appreciated until they are dead, Mendel's revelations would live on, initially in the identification of chromosomes and genes and later in the discovery and sequencing of DNA. At every step of the way, though, one fundamental idea persisted: Who we are is an unwaveringly predictable matter of the genes we've inherited from previous generations.

Mendel called the laws he discovered *inheritance*,[4] and over the years that's how we've come to think of our genetic legacy—as somewhat binary instructions passed from one generation to the next, like a time-worn family heirloom an inheritor doesn't always want but can't throw away.

Or like Ralph's tragic genetic legacy. So why, indeed, was Ralph deviating from Mendel's pea and not showing any visible signs of being affected when so many of his offspring obviously were?

THE GENETIC condition that was burning through Ralph's bloodline follows an autosomal dominant pattern of inheritance. That means you only need one gene with a mutation to be affected with a specific disease. And if you did indeed inherit an offending gene, then chances are generally 50-50 that you'll pass it along to each child you sire. The way we've long understood Mendel's laws of inheritance suggests that if you were unlucky enough to receive a mutated gene that follows this type of inheritance pattern, you'd show the same signs of the disease.

That's probably the genetics you learned about in school, where mapping out pedigree charts made it all too easy, and frankly alluring, to believe we know what we're talking about when it comes to the microscopic molecular magic that makes us who we are. It got a bit more complicated over time, of course, but it all started with an idea, which soon became dogma, that genes come in pairs, and when one gene is dominant over another, it can prompt the same specific trait. Everything from brown eyes to the ability to roll your tongue, grow hair on the back of your fingers, and have detached earlobes were all seen as the result of dominant genes dominating. And correspondingly, it was thought, when two recessive genes pair up they would produce less likely traits such as blue eyes or a hitchhiker's thumb.

But if genetic inheritance always works that way, how is it that Ralph—and all the people who saw him, day in and day out, at the various clinics where he donated his sperm—had no idea that he had such a life-altering disease? Because Mendel, for all he gave to science, missed something of vital importance: variable genetic expressivity.*

Like many other inherited conditions, neurofibromatosis type 1 articulates itself in all sorts of different ways, and sometimes so mildly that it's not recognizable. That's why no one—apparently not even Ralph himself—knew the terrible secret.

Ralph's condition remained hidden because of variable expressivity. This is the reason the same genes can change our lives in very different ways. Identical genes don't always behave identically in different people—even people with completely identical DNA.

Take Adam and Neil Pearson, for example. Born as monozygotic, or identical, twins, these brothers are thought to carry indistinguishable genomes, including a genetic change that causes neurofibromatosis type 1. But Adam has a face that is bloated and disfigured—so badly that a drunken nightclub patron once tried to rip it off, thinking it was a mask. Neil, on the other hand, could pass for Tom Cruise from a certain angle but suffers from memory loss and occasional seizures.[5]

Identical genes, completely different expression. So all of those

---

* Variable expressivity is a measurement of the extent or degree to which one is affected by a genetic mutation or condition.

physical signs I walked you through in chapter 1? They are common expressions and generally indicative of certain genetic conditions, but those traits certainly don't encompass the spectrum of *all* expressions of those genetic conditions.

All of which prompts us to ask, why the difference in expression? Because our genes do not respond to our lives in a binary fashion. As we will come to learn, and contrary to Mendel's findings, even if our inherited genes seem set in stone, the way they express themselves can be anything but. Whereas our inheritance may have been initially understood through a black-and-white Mendelian lens, today we're starting to understand the power of seeing things in full and genetically expressive color.

Which is why now, as physicians, we have a new challenge. Patients look to us to have the answers in clean, discrete categories: benign or malignant, treatable or terminal. The hard part of explaining genetics to patients is that everything we thought we knew is not always static or binary. Figuring out the best way to explain that to patients has become much more critical, since they need the best information possible to help them make some of the most important decisions of their lives.

Because your behavior *can* and *does* dictate your genetic destiny.

WHICH IS why I now want to talk to you about Kevin.

He was in his twenties. Tall and healthy. Handsome, charming, and smart. If I'd known someone at the time who was looking for an eligible bachelor—and if it wouldn't have been an egregious breach of ethics—I might have tried to set him up.

Maybe it was because we were about the same age and came from similar backgrounds. Or maybe it was because we were both involved in health care at the time—he on the eastern and I on the western end of the medical spectrum. Whatever it was, we really seemed to connect.

I met Kevin not too long after his mother passed away from a long and courageous bout with metastatic pancreatic neuroendocrine tumors. Before she died, an astute oncologist suggested genetic testing—and that, in turn, revealed a mutation that sat smack in the middle of her von Hippel-Lindau tumor suppressor gene.

Von Hippel-Lindau syndrome, or VHL, is a genetic condition that predisposes people to tumors and malignancies, including those in the brain, eye, inner ear, kidney, and pancreas. Some researchers have suggested that the infamous Hatfield-McCoy feud may have developed, in part, because of VHL, since many McCoy descendants contemporaneously suffer from tumors of the adrenal gland, which can result in bad tempers.[6] Of course, not everyone with VHL has that sort of symptom—another example of variable expressivity.

And just like the mutated gene that causes NF1 that Ralph was passing along, the gene that causes VHL is inherited in an autosomal dominant fashion. Which means that you only need to get one misbehaving copy from your parents to be affected. Because VHL is an autosomal dominant disorder, we knew that Kevin had a 50-50 chance of inheriting the problem gene from his mother. That was enough to convince him to get checked for the same mutation, which it turns out he had indeed inherited.

There's no cure for VHL, but once we know someone has it, we can increase surveillance for tumors before they become symptomatic. That was what I'd assumed would be the case for Kevin. At least to start out, most people who inherit a mutated or deleted *VHL* gene can still rely on the other working copy to keep cell growth at bay and prevent tumors and malignancies from forming.

We call this the *Knudson hypothesis*, where two or more changes to our genes can set the stage for us to develop cancer. Knowing that you're one gene away from cancer, as Kevin discovered through genetic testing, should make you more careful about how you treat your genes. Radiation, organic solvents, heavy metals, and exposure to plant and fungal toxins are just a few ways to damage and adversely *change* your genes.

The problem is that because VHL can express itself in so many different ways throughout the course of an affected person's life, we never know where and when it's going to pop up. That means we have to keep tabs on just about everything. This entails a regimen of screening and treatment from a team of doctors and allied healthcare workers that will last for the rest of a patient's life.

Not surprisingly, Kevin wanted to know what he could expect moving forward, but because VHL expresses itself in so many different ways, I found it very difficult to answer that question, other than to reiterate the monitoring regimen and which types of tumors and malignancies he'd be at greatest risk for.

"So what you're telling me," he said, "is that we don't know what I'll die from."

"There are treatments for many of the tumors that VHL causes, especially if they're caught early," I replied. "We don't know that you'll die from VHL at all."

"Everybody dies." Kevin chuckled.

I blushed. "Of course. But with treatment—"

"For the rest of my life."

"Yes, that's likely, but—"

"Appointments and checkups, all the time. The stress of constant monitoring. Blood work. Never knowing—"

"Yes, it's a lot, but the alternative—"

"There are always lots of alternatives," he said with a smile, and with that I could see that he'd made his choice.

I was deeply saddened when a few years later he was found to have clear cell metastatic renal carcinoma, a form of kidney cancer. Once again, he resisted any conventional treatment, and he passed away shortly thereafter.

You might be wondering how this is an example of variable expressivity. After all, Kevin died prematurely and tragically, just like his mother. But Kevin died of a different type of cancer and at an earlier age than his mother, so variable expressivity does unfortunately sometimes entail genes behaving in different ways than in the previous or same generation. Using medical surveillance techniques applied by his medical team to keep tabs on his body, Kevin could have used the time after his diagnosis to initiate earlier treatment for his type of kidney cancer. But he chose not to. Given his genetic inheritance, if Kevin simply had asked what types of imaging surveillance his condition required, and then followed through with them, he might not have died prematurely. When it comes to our own health and lives, these choices are ours to make. Our flexible genetic destiny is in many ways ours to determine, if we know what questions to ask and what to do with the answers.[7]

TO BETTER understand the conceptual basis of our flexible inheritance, let's take a quick jaunt to the Jean Remy Library in Nantes, France. That's where, just a few years ago, a librarian sifting through some old files came upon a long-forgotten scrap of sheet music.

The paper was brittle and yellow. The ink had faded into the ancient pulp. But the notations were still clear. The melody was still there. And so it didn't take long for researchers to determine that this little piece of paper—filed away and forgotten for more than a century in the library's archives—was the genuine and exceedingly rare product of Wolfgang Amadeus Mozart's own hand.[8]

Like all of Mozart's more than 600 known works, the melody, several bars in D major thought to have been written a few years before the composer's death, is a set of instructions from the classical composer to all musicians that transcends the centuries. Mozart, it seems, was a fan of the appoggiatura—the sort of brief, dissonant note resolving to a main note that gives Adele's heartrending ballad, "Someone Like You," its peculiar despondent charm.[9] Though most modern composers would use a sixteenth note instead of an appoggiatura, that's nothing more than a small step of musical evolution. And so pianists like Ulrich Leisinger, the director of research at the Mozarteum Foundation in Salzburg, Austria, can use the script to resurrect the long-lost tune. And Leisinger, lucky son of a gun that he is, can do it on the very same 61-key piano upon which Mozart composed many of his concertos more than 220 years ago.[10]

When played, the song crosses the span of space and time like Dr. Who's rickety old time-traveling police box, materializing in the modern world with a mischievous flourish. To Leisinger's trained ear, the tune that emerges when the notes are played is clearly a credo—a liturgical melody. That makes it something of a message in a bottle, because although Mozart wrote a lot of religious music in his younger years, some scholars have questioned whether faith played much, if any, of a role in his latter days.

From the handwriting and the paper, researchers have concluded the score was written around 1787, a time when Mozart—then enjoying steady work on the opera-writing circuit—had no financial need to write church songs. Leisinger believes this reveals Mozart did have an active interest in theology late in his life.

All that from a few dozen notes.

That is roughly how we have long understood DNA. In the same way that modern musicians can read Mozart's instructions and carry them out with near-flawless fidelity, revealing the complexities hidden within, we expect our genetic legacy to be a score upon which is written the music of our lives. And that's true, to some extent.

But it's not the whole story. We are now awakening to a new understanding of our genetic selves and even our evolutionary lineage. Far from being enslaved to a destiny encoded within our DNA, like an obsolete iPod eternally stuck on a requiem, we are learning that there is considerable flexibility within all of us. An inborn ability to change tunes, play our music differently, and, in doing so, overcome some of our previous understanding of our somewhat binary Mendelian genetic destiny.

That's because life, and the genetics that support it, is not like a tattered piece of paper but rather a dimly lit jazz club. Perhaps it is like the Jazzamba Lounge at the Taitu Hotel, in the throbbing center of Ethiopia's capital city, Addis Ababa, where men and women from every corner of the earth come to drink and smoke and laugh and lust.

Just listen:

Clinking glasses. Shuffling chairs. Murmuring voices.

And then, from the shadowy stage, a base:

Baum-baum-baum bada baum-baum bada.

Then the gentle whispers of a brushed snare:

Sha-sssss sha-sssss sha-sssss—sha-sha-sssss.

A cup-muted old trumpet:

Braaaght bra-der-dah braaaght-der-der-bra-dah.

And finally, a sultry female singer:

Oooooo-yah bada baaaaaagh. Hayah hayah hayah bada-yagha.

Just a basic bass line—and all the majesty and tragedy of life to layer upon it.

Now, it's true that for us to cross through the sea of developmental milestones and into adulthood, we do need a significant degree of sophisticated genetic orchestration. So we all start with a score. Older than Mozart. Some of the notes are as old as life on Earth.

But there is plenty of room for improvisation built into our lives. Timing. Timbre. Tone. Volume. Dynamics. Through tiny chemical processes, your body is using each gene you carry like a musician uses an instrument. It can be played loudly or softly. It can be played quickly or slowly. And it can even be played in different ways, as needed, in much the way that the incomparable Yo-Yo Ma can make his 1712 Stradivarius cello play everything from Brahms to bluegrass.

That's genetic *expression*.

Way down, deep and small inside of us, we're all doing that very same thing, churning through the tiny doses of biological energy it takes to change the way our genes express themselves in response to the demands of our lives. And just like musicians who let the culmination of their life experiences and current circumstances affect the way they play their instruments, our cells are guided—expressed—by what has been done and is being done to them at every given moment.

Consider that, and then let's try a little experiment: Stretch out a bit. Move your body. Get comfortable. Now try focusing on your breath. Breathe in, and breathe out. And after a few breaths, tell yourself out loud (or at least in a whisper) that what you do in the world has great value to you and those around you. And now experience how empowering—or just plain silly—this all feels.

There. Right now, inside your body, your genes are at work responding to what you just did, from the moment you began to stretch. Conscious movement is caused by signals sent from your brain, through your nervous system, down to your firing lower motor neurons and all the way to your muscle fibers. Inside those fibers, proteins called actin and myosin are sharing a biochemical kiss, converting chemical energy into mechanical work. And with that, your genes must set to work restoring the chemical ingredients that are necessary every time your brain orders up an action or series of actions—from pressing the volume button on the remote control to running an ultramarathon.

Your thoughts, too, are constantly impacting your genes, which must shift and change, over time, to align your cellular machinery with the expectations you've set and experiences you've had. You're creating memories. Emotions. Anticipation. All of that gets encoded, like an annotation in the margin of an old book, within all of our cells. The hundreds of trillions of synapses in your brain that make this happen are all simply junctions between neurons and cells, and the signals used to communicate must be replaced over time and fed with minuscule doses of chemicals created by your body. And many of our neurons are on the lookout to make new connections as well as maintaining ones that are decades old.

This all happens in response to the demands of your life.

And all of that changes you. Maybe it's just the difference between an appoggiatura and a sixteenth note. Maybe it's even more negligible than that.

But through the flexibility of expression, your life has just changed genetic tunes.

Feeling special? You should. But hold on to your humility, too. Because as we're about to see, these sorts of changes can be seen in all sorts of life forms, large and small. Nor is it just living creatures that can modulate how they respond to life's challenges. Many corporations have employed exactly the same strategies to control their markets or modulate their productions.

As we're about to see, some of these strategies were drawn up long before you were born, and yet they still come into play every time someone gets down on one knee. It's time for me to propose another way to understand the flexibility of genetic expression.

IF YOU'RE in the market for your first shiny rock, or if you're looking for an upgrade, then you might like to know a little secret about the diamond business: Unlike a lot of other types of gemstones, diamonds are actually not all that rare.

It's true. There are lots of diamonds. Lots and lots of them. Big ones. Small ones. Blue, pink, and black. They're mined in dozens of countries and on every continent except Antarctica, although Australian researchers have recently reported finding kimberlite, a type of volcanic rock often rich in diamonds, near the South Pole, so perhaps it's only a matter of time.[11]

Now, if you've ever spent a few paychecks on a diamond, and if you know anything about supply and demand, this might not make a lot of sense. If there are so many diamonds out there, after all, why are they so expensive?

You can thank De Beers for that.

The controversial company, which was founded in 1888 and is headquartered in the grand duchy of Luxembourg, has one of the largest inventories of sparkling stones in the world—most of which have been stashed away. Controlling the process from mining to production to processing to manufacturing, De Beers maintained a near worldwide monopoly on the diamond exchange for generations, releasing just the right amount of product into the market at just the right time so as to keep prices up and the market stable—and ensuring that a relatively common rock remained precious in the eyes (and wallets) of its beholders.[12]

Savvy marketing tricks have done the rest. Before World War II, very few people exchanged engagement rings—and diamonds were just one of many kinds of stones that might be on them. But in 1938, De Beers hired a Madison Avenue adman by the name of Gerold

Lauck to figure out how to persuade young men that a shiny piece of well-compressed carbon was the only way to express an intention of betrothal to a prospective mate. And by the early 1940s, Lauck's marketing magic had managed to convince a good segment of the Western world that diamonds are indeed a girl's best friend.[13]

The industrialist Henry Ford would have loved to corner the market in that way. He certainly may have conspired to do so, but Ford's product and its production were so complicated at the time that he had no choice but to deal with a lot of suppliers.

That frustrated Ford to no end. The People's Tycoon, as he was known, was perhaps the world's first famous disciple of industrial efficiency—which we now can understand is rooted in many of the same strategies exploited by our genomes through genetic expression. Not surprisingly, Ford spent a lot of time working on streamlining the process as much as possible.

"We have found in buying materials that it is not worthwhile to buy for other than immediate needs," Ford wrote in his 1922 book, *My Life and Work*. "We buy only enough to fit into the plan of production, taking into consideration the state of transportation at the time."[14]

Alas, Ford lamented, the state of transportation was far from perfect. If it were perfect, he said, " . . . it would not be necessary to carry any stock whatsoever. The carloads of raw materials would arrive on schedule and in the planned order and amounts, and go from the railway cars into production. That would save a great deal of money, for it would give a very rapid turnover and thus decrease the amount of money tied up in materials."

Ford's words were prescient, but he went to his grave without solving this problem. Eventually, Japanese car manufacturers were responsible for making great leaps in a system of production that tied the supply chains to immediate demand, a process we now know as just in time, or JIT, production. The business lore is that executives from Toyota were exposed to JIT while in the United States in the 1950s—not by the American car companies that they were visiting at the time, but rather during a side trip to the first self-service grocery store called the Piggly Wiggly. One of the grocery chain's novel approaches was to have inventory automatically restocked as soon as it was removed from the shelf.[15]

There are many benefits to employing this type of technique—chiefly, when done right, it can be a major moneymaker and saver. Of course, it's not without its risks, one of the biggest being that the entire process becomes susceptible to supply shocks, events such as natural disasters or worker strikes that can interrupt delivery of raw materials and, in doing so, leave a factory completely idle and customers empty-handed.

Apple experienced another of the drawbacks associated with JIT manufacturing: An unprecedented wave of demand for iPad Minis almost drowned out the company's ability to produce the product when it couldn't get the materials to make them into its factories fast enough.

Understanding how businesses employ certain strategies similar to genetic expression can help us understand the biological strategies used by most of our cells to keep the costs of living down. Just like corporations, our bodies employ an unforgiving bottom line. Doing this makes the likelihood of our continued existence possible.

And in that regard, we employ more of a Costco model of operations than a Walmart one. Since there's a biological cost every time we use our genes to make anything, we aim to get the most out of what we make. Just like Costco does with its employees, our biology is configured for higher labor productivity—meaning we aim to have the smallest number of enzyme employees for the jobs we need to get done. Enzymes behave like microscopic molecular machines and are an example of structures that are coded for by our genes. Some enzymes are able to speed up chemical processes, while others like pepsinogen, when activated, help us digest our proteinaceous meals. Other enzymes, such as those that belong to the P450 family, detoxify poisons we may be knowingly or unwittingly consuming.

We generally only produce what we need when we need it and try to keep what we store to a minimum. And we do that through genetic expression.

Just like diamonds, which take millions of years and a whole lot of pressure to create, enzymes are biologically expensive to produce. To mitigate the cost of production, many of our enzymes can be induced. Which means that, when we need certain enzymes, our bodies can marshal up more of their resources to produce more of them on call, churning out the biological equivalent of iPad Minis to meet an increased demand. You may have inherited the genes for an enzyme, but that doesn't always ensure that your body will use it.

There's a good chance you've experienced this at some point in your life, unaware of your active role in the process. If you've ever binged on alcohol—over a long holiday weekend, perhaps—then you've been there. In response to your partying, your liver cells worked overtime to make all the enzymes they needed to deal with that unexpected deluge of margaritas.

The means of increasing production to meet demand—in this case alcohol dehydrogenase to break down ethanol—are always there, latent in your liver cells, ready for your next binge. But it may not be stored in large quantities, because just like extra parts sitting around on the factory floor, enzymes not only take up space but are costly to produce and maintain when you're not drinking to excess.

Almost all of the biological world is driven in the same way to streamline the cost of living. And it needs to be. Spend all your energy on enzymes you're not going to use, and you'd be diverting precious resources from other daily concerns, like the continual process of brain plasticity.

Astronauts provide a great example of this. Soon after arriving on the International Space Station, their hearts shrink by as much as a quarter of their original size.[16]

In the same way that trading in a supercharged 300-horsepower Ford Mustang for a Mini Cooper with less than half the ponies would save you a whole lot of money at the gas station, the weightless environment of space means astronauts don't need as big a cardiac engine.* But that's also why, upon returning to Earth and reexperiencing gravity, space travelers often become light-headed and sometimes black out: Their hearts, like a Mini trying to scale a steep mountain road, just can't push enough blood—and the crucial oxygen it carries—up to the brain.

You don't need to travel up to the space station for your heart to shrink. Just a few weeks in bed are all you need for it to begin to atrophy.[17] But our bodies are also quite amazing at recovery—we just have to convince them that we need the power. And that's not always a tough sell, because our cells are incredibly malleable. What we do every day makes a big difference in what our genes tell them to do—which is just another genetic motivation for you to get off the couch.

Before we leave genetic expression, there's just one more thing I'd like us to explore together.

AT FIRST glance, *Ranunculus flabellaris* might not seem like that big a deal. The yellow water buttercup, which grows prolifically in

---

* Our hearts use a lot of energy to move our blood against the pull of gravity. If we're in orbit, our blood becomes weightless, and then we can get the same degree of circulation with a lot less force. Which is why in space we can get away with having a much smaller heart.

forested wetlands in the United States and southern Canada, isn't much to look at. Yet what you're looking at, when you find one, is a plant that can completely change its appearance depending on how close it finds itself to water, a behavior we call heterophylly.

The buttercup usually grows along a riverbank, which can be a precarious place for a plant to be as rivers are prone to overflow from season to season. That could be deadly for a delicate little flower like this one, but living on the edge of this habitat doesn't deter this plant. Rather, it allows it to thrive, because genetic expression gives it the ability to completely change the shape of its leaves—from rounded blades to threadlike hairs that can float if the river spills over its banks.[18]

When this change happens, the buttercup's genome stays the same. To a passerby, it might look like a completely different plant, but deep inside, its genes haven't changed. Only its expressed phenotype, or observable appearance, is altered.

And just as an astronaut's body can go from Mustang to Mini Cooper and back again, based on the conditions in which they live, another change in environment for the buttercup—the height of the river decreasing with the changing season—switches the plant back to the previous type of leaf growth. It's all a matter of survival.

Expression is just one of many strategies that plants, insects, animals, and even humans employ to deal with the rigors of life. In all of them, though, one thing is key: flexibility.

What we are now learning is that our genes are part of a larger flexible network. This is contrary to much of what we've been told about our genetic selves. Our genes aren't as fixed and rigid as most of us have been led to believe. If they were, we wouldn't be able to adjust—just like the yellow water buttercup does—to the ever-changing demands of our lives.

The thing Mendel couldn't see in his peas—and that generations of geneticists continued to miss after his death—is that it's not only what our genes give to us that's important, but also what we give to our genes. Because, as it turns out, nurture can and does trump nature.

And as we are about to see, it happens all the time.

CHAPTER 3

Changing Our Genes

How Trauma, Bullying, and Royal Jelly Alter Our Genetic Destiny

Most people know about Mendel's work with peas. Some have heard of his truncated work with mice. But what most people don't know is that Mendel also worked with honeybees—which he called "my dearest little animals."

Who can blame him for such adulation? Bees are endlessly fascinating and beautiful creatures—and they can tell us a lot about ourselves. For instance, have you ever been witness to the awesome and fearsome sight of an entire colony of bees swarming and on the move? Somewhere in the middle of that ethereal tornado is a queen bee that has left the hive.

Who is she to deserve such a grand parade?

Well, just look at her. For starters, just like human fashion models, queens have longer bodies and legs than their sister workers. They're more slender and have smooth, rather than furry, abdomens. Because they often need to protect themselves from entomological coups from younger royal upstarts, queen bees have stingers that can be reused on demand, unlike female worker bees, who die after using their stingers just once. Queen bees can live for years, though some of their workers live only a few weeks. They can also lay thousands of eggs in a day, while all their royal needs are tended to by sterile workers.*

So yeah, she's kind of a big deal.

Given the incredible differences between them, you could easily assume that queens differ genetically from the workers. That would make sense—after all, their physical traits differ considerably from their sister worker bees. But look deeper—DNA deeper—and a very different story emerges. The truth is that, genetically speaking, the queen is nobody special. A queen bee and her female workers can come from the same parents, and they can have completely identical DNA. Yet their behavioral, physiological, and anatomical differences are profound.

Why? Because larval queens eat better.

That's it. That's all. The food they eat changes their genetic expression—in this case through specific genes being turned off or on, a mechanism we call epigenetics. When the colony decides it's time for a new queen, they choose a few lucky larvae and bathe them in royal jelly, a protein- and amino acid–rich secretion produced by glands in the mouths of young worker bees. Initially, all larvae get a taste of royal jelly, but workers are quickly weaned. The little princesses, however, eat and eat and eat until they emerge as a blue-blooded brood of elegant empresses. The one who murders all the rest of her royal sisters first gets to be queen.

Her genes are no different. But her genetic expression? Royal.[1]

Beekeepers have known for centuries—maybe longer—that larvae bathed in royal jelly will produce queens. But until the genome for the western honeybee, *Apis mellifera*, was sequenced in 2006, and the

---

* Worker honeybees, at times, can lay eggs that will hatch into drones (male bees). But given the complexities of their reproductive genetics, worker bees are incapable of laying eggs that will become other female workers.

specific details of caste differentiation were worked out in 2011, no one knew exactly why.

Like every other creature on this planet, bees share a lot of genetic sequences with other animals—even us. And researchers quickly noticed that one of these shared codes was for DNA methyltransferase, or Dnmt3, which in mammals can change the expression of certain genes through epigenetic mechanisms.

When researchers used chemicals to shut down the Dnmt3 in hundreds of larvae, they got an entire brood of queens. When they turned it back on in another batch of larvae, they all grew to be workers. So rather than having something more than their workers, as might be expected, queens actually have a little less—the royal jelly the queens eat so much of, it appears, just turns down the volume on the gene that makes honeybees into workers.[2]

Our diet differs from that of bees, of course, but they (and the clever researchers who study them) have given us lots of amazing examples of how our genes express themselves to meet the demands of our lives.[3]

Like humans who fill a series of set roles during their lives—from students to workers to community elders—worker bees also follow a predictable pattern from birth to death. They start as housekeepers and undertakers, keeping the hive clean and, when necessary, disposing of their dead siblings to protect the colony from disease. Most then become nurses, working together to keep tabs on each larval member of the hive more than a thousand times a day. And then, right around the ripe old age of two weeks, they set off to forage for nectar.

A team of scientists from Johns Hopkins University and Arizona State University knew that sometimes, when more nurse bees are needed, foraging bees will go back to do that job. The scientists wanted to know why. So they looked for differences in gene expression, which can be found by searching for chemical "tags" that rest atop certain genes. And indeed, when they compared the nurses with the foragers, those markers were in different places on more than 150 genes.

So they played a little trick. When the foragers were off searching for nectar, the researchers removed the nurses. Not willing to permit their young ones to be neglected, upon their return the forager bees immediately reverted to nurse duties. And just as immediately, their genetic tagging pattern changed.[4]

Genes that weren't being expressed before, now were. Genes that were, now weren't. The foragers weren't just doing another job—they were fulfilling a different genetic destiny.

Now, we might not look like bees. And we might not feel like bees. But we share a striking number of genetic similarities with bees, including Dnmt3.[5]

And just like those bees, our lives can be momentously impacted by genetic expression, for better or for worse.

Take spinach, for instance. Its leaves are rich in a chemical compound called betaine. In nature or on a farm, betaine helps plants deal with environmental stress, such as low water, high salinity, or extreme temperatures. In your body, though, betaine can behave as a methyl donor—part of a chain of chemical events that leaves a mark on your genetic code. And researchers at Oregon State University have found that, in many people who eat spinach, the epigenetic changes can help influence how their cells fight back against genetic mutations caused by a carcinogen in cooked meat. In fact, in tests involving laboratory animals, researchers were able to cut the incidence of colon tumors nearly in half.[6]

In a very small but important way, compounds within spinach can instruct the cells within our bodies to behave differently—just like royal jelly instructs bees to develop in different ways. So yes, eating spinach seems to be able to change the expression of your genes themselves.

REMEMBER WHEN I told you that Mendel, if Bishop Schaffgotsch had not curtailed his work with mice, might have stumbled upon something even more revolutionary than his theory of inheritance? Well, now I'd like to tell you about how that idea finally came to light.

First of all, it took time. More than 90 years had passed since Mendel's death when, in 1975, geneticists Arthur Riggs and Robin Holliday, working separately in the United States and Great Britain, respectively, almost simultaneously came upon the idea that, while genes were indeed fixed, they could perhaps be expressed differently in response to an array of stimuli, thus producing a range of traits rather than the fixed characteristics commonly thought to be associated with genetic inheritance.

Suddenly, the idea that the way genes are inherited could only be changed by the epically slow process of mutation was thrown into immediate dispute. But just as Mendel's ideas had been roundly ignored, so too were the theories being offered by Riggs and Holliday. Once again, an idea about genetics that was ahead of its time failed to gain traction.

It would be another quarter century before these ideas—and their profound implications—would gain broader acceptance. And that came as the result of the striking work of a cherub-faced scientist named Randy Jirtle. Like Mendel, Jirtle suspected that there was more to inheritance than met the eye. And, like Mendel, Jirtle suspected the answers could be found in mice.

Experimenting with agouti mice, which carry a gene that renders them plump and bright orange like a Muppet, Jirtle and his associates at Duke University came upon a discovery that, at the time, was simply stunning. By doing nothing more than changing the diet of females by the addition of a few nutrients such as choline, vitamin B12, and folic acid, starting just before conception, their offspring would be smaller, mottled brown, and altogether more mouse-like in appearance. Researchers would later discover that these mice were less susceptible to cancer and diabetes as well.

Same exact DNA. Completely different creature. And the difference was simply a matter of expression. In essence, a change in the mother's diet tagged her offspring's genetic code with a signal to turn off the agouti gene, and that turned-off gene then became inherited and was passed down across generations.

But that's just the beginning. In the fast-paced world of twenty-first-century genetics, Jirtle's Muppets have already been relegated to syndicated reruns. Every day we're learning new ways to alter genetic expression—in the genes of mice and men. The question isn't whether we can intervene; that's now a given. Now we're examining how to do it with new drugs that are already approved for human use, in ways that will hopefully result in longer and healthier lives for ourselves and for our children. What Riggs and Holliday theorized about—and what Jirtle and his colleagues brought into popular acceptance—is now known as epigenetics. Broadly, epigenetics is the study of changes in gene expression that result from life conditions, such as those seen in honeybee larvae that are doused in royal jelly, without changes in the underlying DNA. One of the fastest growing and most exciting areas of epigenetic study is its heritability, the investigation of how these changes can impact the next generation, and every generation down the line.

ONE COMMON way changes in genetic expression occur is through an epigenetic process called methylation. There are many different ways in which DNA can be modified without the underlying string of nucleotide letters being altered. Methylation works by the use of a chemical compound, in the shape of three-leaf clovers made up of hydrogen and carbon, that is attached to DNA that alters the genetic structure in such a way as to program our cells to be what they're supposed to be and to do what they're supposed to do—or what they've been told to do by previous generations. Methylation "tags" that turn genes on and off can give us cancer, diabetes, and birth defects. But don't despair—because they can also affect gene expression to give us better health and longevity.

And such epigenetic changes seem to have consequences in some unexpected places. For instance, at a summer weight-loss camp.

Genetic researchers decided to follow a group of 200 Spanish teens who were on a 10-week quest to battle the bulge. What geneticists discovered was that they could actually reverse engineer the campers' summer experience and predict which of the teens would lose the most weight depending on the pattern of methylation—the way their genes were turned off or on—in around five sites in their genome before summer camp even began.[7] Some kids were epigenetically primed to lose the bulge at summer camp while others were going to keep it on, despite diligent adherence to their counselors' dietary protocol.

We're now learning how to apply the knowledge gained from studies such as these to capitalize on our own unique epigenetic makeup. What the teenagers' methylation tags teach us is how critical it is to get to know our own distinctive epigenome in matters of weight loss, and so much else. Learning from these Spanish summer campers, we can start to mine our epigenome to find the information we need for the most optimal weight-loss strategies. For some of us that may mean saving on the exorbitant fees of a summer weight-loss adventure that is destined not to work.

But far from being static, our epigenome, along with the DNA that we've inherited, can also be impacted by what we do to our genes. We're quickly learning that epigenetic modifications, like methylation, are remarkably easy to impact. In recent years geneticists have devised a number of ways to study and even reprogram methylated genes—to turn them on and off, or to crank the volume up or down.

Changing the volume of our genetic expression can mean the difference between a benign growth and a raging malignancy.

These epigenetic changes can be caused by the pills we swallow, the cigarettes we smoke, the drinks we consume, the exercise classes we attend, and the X-rays we undergo.

And we can also do it with stress.

Building on Jirtle's work on agouti mice, scientists in Zurich wanted to see whether early childhood trauma could impact gene expression, so they stole pup mice away from their mothers for three hours, then returned the blind, deaf, and furless little things to their mommies for the rest of the day. The next day, they did it again.

Then, after 14 consecutive days, they stopped. Eventually, as all mice do, the little ones gained sight and hearing, grew some fur, and became adults. But having suffered two weeks of torment, they grew up to become significantly maladjusted little rodents. In particular, they seemed to have trouble evaluating potentially risky places. When put in adverse situations, instead of fighting or figuring it out, they just gave up. And here's the amazing part: They transmitted these behaviors to their own pups—and then to the offspring of their offspring—even if they had no involvement whatsoever in the rearing.[8]

In other words, a trauma in one generation was genetically present two generations down the line. Incredible.

It's definitely worth noting here that the genome for a mouse is about 99 percent similar to ours. And the two genes impacted in the Zurich study—called *Mecp2* and *Crfr2*—are found in mice and people alike.

Of course, we can't be sure that what happens in mice will happen in humans until we do, in fact, see it. That can be challenging to do, because our relatively long lives make it hard to conduct tests that explore generational changes, and when it comes to humans, it's a lot harder to separate nature and nurture.

But that doesn't mean we haven't seen epigenetic changes related to stress in humans. We most certainly have.

REMEMBER WHEN I asked you to go back to the seventh grade? For some of us going back that far might evoke some rather unpleasant memories, events that, given a choice, we'd rather not recall. The real numbers are hard to come by, but it's thought that at least three quarters of all children have been bullied at some point in their lives, which means there's a good chance you were on the receiving end of such unfortunate experiences yourself while you were growing up. And as some of us have become parents since then, the concern for our children's own experiences and safety both at school and beyond has only grown.

Until very recently, we've been thinking and speaking about the serious and long-term ramifications of bullying in predominantly psychological terms. Everyone agrees that bullying can leave very significant mental scars. The immense psychic pain some children and teens experience can even lead them to consider and act on desires to physically harm themselves.

But what if our experiences of being bullied did a lot more than just saddle us with some serious psychological baggage? Well, to answer that question, a group of researchers from the UK and Canada decided to study sets of monozygotic "identical" twins from the age of five. Besides having identical DNA, each twin pair in the study, up until that point, had never been bullied.

You'll be glad to know that these researchers were not allowed to traumatize their subjects, unlike how the Swiss mice were handled. Instead, they let other children do their scientific dirty work.

After patiently waiting for a few years, the scientists revisited the twins where only one of the pair had been bullied. When they dropped back into their lives, they found the following: present now, at the age of 12, was a striking epigenetic difference that was not there when the children were five years old. The researchers found significant changes only in the twin who was bullied. This means, in no uncertain genetic terms, that bullying isn't just risky in terms of self-harming tendencies for youth and adolescents; it actually changes how our genes work and how they shape our lives, and likely what we pass along to future generations.

What does that change look like genetically? Well, on average, in the bullied twin a gene called *SERT* that codes for a protein that helps move the neurotransmitter serotonin into neurons had significantly more DNA methylation in its promoter region. This change is thought to dial down the amount of protein that can be made from the *SERT* gene—meaning the more it's methylated the more it's "turned off."

The reason these findings are significant is that these epigenetic changes are thought to be able to persist throughout our lives. This means that even if you can't remember the details of being bullied, your genes certainly do.

But that's not all these researchers found. They also wanted to see if there were any psychological changes between the twins to go along with the genetic ones that they observed. To test that, they subjected the twins to certain types of situational testing, which included public speaking and mental arithmetic—experiences most of us find stressful and would rather avoid. They discovered that one of the twins, the one with a history of being bullied (with a corresponding epigenetic change), had a much lower cortisol response when exposed to those unpleasant situations. Bullying not only turned those children's *SERT* gene to low, it also turned down their levels of cortisol when stressed.

At first this may sound counterintuitive. Cortisol is known as the "stress" hormone and is normally elevated in people under stress. Why, then, would it be blunted in the twin who had a history of being bullied? Wouldn't you think they would be *more* stressed in a heightened situation?

This gets a little complicated, but hang tight: As a response to the persistent bullying trauma, the *SERT* gene of the bullied twin can alter the hypothalamic-pituitary-adrenal (HPA) axis, which normally helps us cope with the stresses and tumbles of daily living. And according to the scientists' findings in the bullied twin, the greater the degree of methylation, the more the *SERT* gene is turned off. The more it's turned off, the more blunted the cortisol response. To understand the sheer depth of this genetic reaction, this type of blunted cortisol response is also often found in people with post-traumatic stress disorder (PTSD).

A spike of cortisol can help us through a tough situation. But having too much cortisol, for too long, can short-circuit our physiology pretty quickly. So, having a blunted cortisol response to stress was the twin's epigenetic reaction to be being bullied day after day. In other words, the twin's epigenome changed in response to protect them from too much sustained cortisol. This compromise is a beneficial epigenetic adaptation in these children that helps them survive persistent bullying. The implications of this are nothing short of staggering.

Many of our genetic responses to our lives work in such a fashion, favoring the short over the long term. Sure, it's easier in the short term to dull our response to persistent stress, but in the long run, epigenetic changes that cause long-term blunted cortisol responses can cause serious psychiatric conditions such as depression and alcoholism. And not to scare you too much, but those epigenetic changes are likely heritable from one generation to the next.

If we're finding such changes in individuals like the bullied twin, then what about traumatic events that affect large swaths of the population?

IT ALL started, tragically, on a crisp and clear Tuesday morning in New York City. More than 2,600 people died in and around New York's World Trade Center on September 11, 2001. Many New Yorkers who were in direct proximity to the attacks were traumatized to the point of suffering from post-traumatic stress disorder in the months and years to come.

And for Rachel Yehuda, a professor of psychiatry and neuroscience at the Traumatic Stress Studies Division at the Mount Sinai Medical Center in New York, the terrible tragedy presented a unique scientific opportunity.

Yehuda had long known that people with PTSD often had lower levels of the stress hormone cortisol in their systems—she'd first seen that effect in combat veterans she studied in the late 1980s. So she knew where to start when she began looking at samples of saliva collected from women who were at or near the Twin Towers on 9/11, and who were pregnant at that time.

Indeed, the women who ultimately developed PTSD had significantly lower levels of cortisol. And so did their babies after birth—especially the ones who were in the third trimester of development when the attacks occurred.

Those babies are older now, and Yehuda and her colleagues are still investigating how they've been impacted by the attacks. And they've already established that the children of the traumatized mothers are likely to become distressed more easily than others.[9]

What does all this mean? Taken together with the animal data we now have, it is safe to conclude that our genes do not forget our experiences, even long after we've sought therapy and feel that we've moved on. Our genes will still register and maintain that trauma.

And so the compelling question remains: Do we or do we not pass on the trauma we experience, be it bullying or 9/11, to the next generation? We previously thought that almost all of these epigenetic marks or annotations that were made on our genetic code, like those made in the margins of a musical score, were wiped clean and removed before conception. As we prepare to leave Mendel behind, we are now learning that this is likely not the case.

It is also becoming apparent that there are actually windows of epigenetic susceptibility in embryonic development. Within these important time frames, environmental stressors such as poor nutrition affect whether certain genes become turned off and on and then affect our epigenome. That's right, our genetic inheritance becomes imprinted during pivotal moments of our fetal lives.

When exactly those moments occur no one yet knows precisely, so to be safe, moms now have a genetic motivation to watch their diets and stress levels consistently throughout gestation. Research is now even showing that factors such as a mother's obesity during pregnancy can cause a metabolic reprogramming in the baby, which puts the baby at risk for conditions such as diabetes.[10] This further buttresses the growing movement within obstetrics and maternal-fetal medicine that discourages pregnant woman from eating for two.

And, as in the example of the traumatized Swiss mice, we've already seen that many of these epigenetic changes can be passed on from one generation to the next. Which makes me think that the likelihood is rather high that in the coming years we'll have overwhelming evidence that humans are not immune from this type of epigenetic traumatic inheritance.

In the meantime, given the tremendous amount we've learned about what inheritance really means and what we can do to impact our genetic legacy—in good ways (spinach, perhaps) and bad (stress, it would appear)—you are far from helpless. While it may not always be possible to break completely free from your genetic inheritance, the more you learn, the more you will come to understand that the choices you make can result in a big difference in this generation, the next one, and possibly everyone else down the line.

Because what we do know is that we are the genetic culmination of our life experiences, as well as every event our parents and ancestors ever lived through and survived—from the most joyous to the most heartrending. By examining our capacity to change our genetic destiny through the choices we make and then pass those changes along through generations, we are now in the midst of fully challenging our cherished Mendelian beliefs regarding inheritance.

CHAPTER 4

Use It or Lose It

How Our Lives and Genes Conspire to Make and Break Our Bones

Doctors and drug dealers. Those are the only people who seem to carry pagers anymore—and when I check my beeper in a crowded restaurant or before heading into the theater, I often wonder what other people must be thinking.

When it went off one recent morning, I was just approaching the front of a long line at the Starbucks in a bustling hospital atrium. From where I stood, I could have almost grabbed a cup and scribbled my own order on it, but the person in front of me was taking her time ordering a venti double shot, soy mocha something or other.

So close, yet so far.

I stepped away to return the page. The woman on the other end of the line was from the pediatric team that was caring for a young patient with multiple bone fractures. She asked if I could come by for a consultation involving a little girl. They were just finishing up some routine work but would be ready for me in 15 minutes or so. I jotted the room number down on a napkin and got back into a line that had grown significantly longer in the two minutes since I'd stepped away.

I didn't really mind—the extra few minutes in queue gave me

time to collect my thoughts. I began running through an internalized algorithm for recurrent fractures in a young child—*if this, then that...if that, then this*—that would help me evaluate her condition.

And as I did that, I thought about the special connection our bones give us to the rest of our body.

From plastic Halloween yard decorations to *The Pirates of the Caribbean*, we've all had plenty of opportunities to get acquainted with skeletons. Our collective familiarity—even if you can't name a single one of your 206 bones, you can probably draw a very basic map of your skeleton—makes them easy to visualize when it comes to talking about how our bodies respond to the ever-changing demands of our lives.

Like most of our body's systems, our skeleton follows the use-it-or-lose-it dictum of biological life. In response to our actions or inactions, our genes can be called upon to put into motion processes that can give us strong and malleable bones or ones that are porous and brittle as chalk. In this way, our life experiences affect our genes.

But not all of us inherit the genetic know-how to create the types of bones that are needed for the skeletal flexibility that life requires. That's what I surmised might be the case as, hot Earl Grey tea finally in hand, I rode up to the seventh floor and knocked on the door of the patient's room. On the bed before me, with black locks and wearing a tiny hospital gown, was a sweet little three-year-old girl named Grace.

There was perspiration on her brow, likely from the pain she was feeling from her fractures. I made a mental note of that as I dove into the rapid scan that takes place whenever I pull the curtain that provides patients a little extra privacy from the hospital's busy hallways.

Quickly, I focused on one very important feature.

Her eyes.

LIZ AND David couldn't have a biological child of their own. And for a long time, that seemed just fine.

Liz was a gifted graphic artist. David was an accountant with his own company. They were both quite happy to put their time into their careers and focus their attention on each other. On vacation, they traveled around the world. At home, they enjoyed the best of everything.

They'd watched their parenting friends expend immense amounts of energy just coming up with a weekly car-pool plan. There were schools to consider. Parent-teacher conferences to attend. Music classes. Athletic practices. Summer camps. There were 2 a.m. nightmares and 6 a.m. wake-ups. It was all too much.

Which is why they themselves were surprised to discover that one day, seemingly out of the blue, their perspective had changed.

There were children around the world who needed parents. But as Liz studied the tragically imbalanced mortality rates for orphan girls in China, she knew what they needed to do.

The world's most populous nation instituted its one-child policy in 1979, at a time when the nation was about to become the first in the world to cross the one billion population threshold, even as many of its residents struggled to find shelter, food, and work. Government medical authorities issued birth control, but when it failed, abortion became the standard option.* Those who did give birth to a second or sometimes even third child, especially in urban areas, often had no option but to leave those children at the doorstep of a state-run orphanage.

But one parent's sorrow could be another's joy. The Chinese system had created a glut of orphans, especially female ones, more than could be adopted by Chinese couples who couldn't have children of their own. Within five years of implementing the controversial

---

* We'll discuss the unexpected history behind this phenomenon further in chapter 10.

policy, a nation that had little history of permitting children to go overseas for adoption had become a key "sending" country.

And by 2000, China had become the single-largest foreign provider of adoptive children to U.S. and Canadian families. Although the numbers have waned somewhat in recent years, China remains one of the most significant contributors to the adoptive pool for North American parents.

Liz and David understood that this path would be full of challenges, too. The process has, at times, been marred by corruption. And even when done right—from the moment prospective parents begin working with an agency to the moment they bring a child home—it can take years. But couples willing to adopt children who have some type of physical problem—generally medically "correctable" issues such as a cleft lip—are sometimes treated to a bit of bureaucratic wheel-greasing.

One such condition is called congenital hip dysplasia, a fairly common disorder in which children are born with a hip that easily dislocates. In most developed countries where children have good access to health care, hip dysplasia cases are generally treatable if corrected early in life. But in countries that lack medical resources, these children can end up having significant handicaps. That, the would-be parents were told, was Grace's problem.

But Liz and David were instantly in love. From the moment they first saw a picture of Grace, they knew she was the girl for them. They gathered Grace's documents from the adoption facilitator and consulted with a pediatrician, who assured them Grace's situation would likely be easy to treat once she arrived in North America.

Giving Grace the medical care she needed seemed a relatively small hurdle to overcome for the honor of becoming her parents. With that, they booked their tickets to China and started childproofing their home.

They didn't know a lot else about their daughter-to-be. What they were told at the time was that Grace had been left at the orphanage doorstep a year earlier and was thought to be two years old. That was about it. When Liz and David arrived at the orphanage in the southwest Chinese city of Kunming to pick up their daughter, they learned there was a lot more.

They'd known to expect a spica, the type of cast that starts at the waist and holds the legs akimbo. The only surprise was how big it was and how small she was—it seemed as though the tiny little girl, weighing just about 12 pounds, had been swallowed by a big plaster monster.

Still, given the assurances they'd received from their doctor, they remained confident that Grace's condition was only temporary and perfectly treatable. When an orphanage worker saw how unbothered they were by the challenges posed by the little girl's condition, she pulled them aside to tell them how excited she was that Grace would be going home with them.

"You are her destiny," she said.

And they absolutely were.

A few days later they were back in North America and, after a quick visit and examination by the pediatrician, they were able to get Grace out of her cast and scheduled for a follow-up visit to begin addressing the hip dysplasia.

But hidden under the cast, the little girl's waist and legs were terribly scrawny. And less than 24 hours after the spica was removed, Grace had broken her left femur and right tibia.

Rather than helping address the hip dysplasia, it seemed at the time that the cast had made things worse, allowing her bones to attenuate to the point of glasslike fragility. Back into a cast she went.

A few months later and finally free of the cast, Grace was resting in her mother's arms in a sporting goods store where they were looking to purchase a canoe for an upcoming camping trip. She shifted her body to point at a pink one that she fancied.

The sound, the little girl's mother would later tell me, was like a gunshot. Liz shuddered. Grace wailed. Minutes later the frantic new mother and screaming toddler were back at the hospital. Grace's leg had broken again.

Even before I began taking the history from her parents, it was clear to me that there was far more at play in Grace's case than congenital hip dysplasia.

The answer was in her eyes. Human eyes are distinct in that the sclera—the so-called "whites of our eyes"—is visible, whereas in most other species' eyes, it's mostly hidden behind folds of skin and the optical socket. For dysmorphologists, that presents an extra window of opportunity to understand what's happening within a patient's genes.

Grace's sclera wasn't white but a light shade of blue—and that, along with her history of bone fractures, told me that she was likely suffering from a type of osteogenesis imperfecta, or OI, a condition in which a genetic defect inhibits the production or quality of collagen, which is essential for strong and healthy bones. The same lack of collagen that was making her bones so brittle was also giving her sclera its slightly blue hue, and a quick peek at her teeth—which were translucent at the tips for the same reason—told me I was on the right track.

It wasn't so long ago that OI might not have been diagnostically considered at all. In the past few years, though, the condition has been getting a lot of attention, thanks in no small part to an unquestionably adorable kid named Robby Novak—better known as Kid President—whose viral string of pep talk videos calling on the world to "stop being boring" have been watched by tens of millions of people around the world.

But Robby, who suffered more than 70 broken bones and underwent 13 surgeries before he was 10, didn't set out to draw attention to OI. "I want everybody to know I'm not that kid who breaks a lot," he told CBS News in the spring of 2013. "I'm just a kid who wants to have fun."[1] Robby's story, though, has inspired many people to take a harder look at OI and what is being done to help those who suffer from it.

The disease has also been in the news for other reasons—mainly because it has become a factor in thousands of child abuse investigations. Take Amy Garland and Paul Crummey, for instance. The British couple was accused by social workers of abusing their young son, who was found to have eight fractures in his arms and legs shortly after he was born. After being arrested on suspicion of abuse, Amy and Paul were also banned from seeing their children without proper supervision. The courts wouldn't take the infant away, since he was still breastfeeding, so they ordered Amy to move into a facility where they could monitor her. In a case of reality imitating reality television, the local authority had them placed in a home where it could watch the family 24 hours a day through closed-circuit cameras, as though they were contestants on the TV show *Big Brother*.[2]

It took 18 months for social workers and others involved to realize they'd made a terrible mistake. Amy and Paul's son wasn't suffering from abuse but from OI.

It's understandable why an X-ray of a child who suffers from OI can look like evidence of child abuse, as such pictures will reveal multiple fractures in different stages of healing. But given cases in which social workers and doctors—seeking only to protect children from danger—have wrongly accused good parents of being abusers, most courts now ask that the possibility of OI be considered as part of abuse investigations.

Although such screening is becoming more widespread, the problem for those involved in cases of suspected abuse is that it can take a while to rule out OI. Despite what you might have been led to believe from police dramas on television, understanding what someone's DNA is telling us is not always as easy as walking into a hospital lab and looking under a microscope. Since there are many ways in which a person can have brittle bones, finding the cause, through biochemical and genetic investigations, can take weeks or even months to resolve. Given increased awareness of the possibility of OI, the relative rarity of the disease (some 400 cases a year in the United States alone), and the apparent epidemic of child abuse (more than 100,000 substantiated cases of physical abuse and some 1,500 deaths each year),[3] many social service and law enforcement agencies still make the heartrending decision to be safe instead of sorry.

Thankfully, Grace's history in no way suggested that abuse should be near the top of the list of possible causes for her multiple fractures. That meant we could immediately focus on what was going wrong, with her new parents as full partners on our quest for answers and interventions that would give Grace the healthy, happy life she deserved.

Not so long ago, there wasn't much we could do for the so-called nonlethal types of OI. Today the condition is still a challenge, but one look at Grace will tell you it's not an insurmountable one.

Of course, no single kind of therapy is usually enough to address the complex issues that emanate from deep inside our genes. But when we begin to piece together the right combination of drugs, physical therapy, and technomedical interventions, we can have a real impact. With those tools—and her own bravery, persistence, and dedicated parents—Grace has grown from a tiny, fragile toddler to a tough and adventurous little girl. With every new step she takes, her life experiences shape and defy her very genetic code. Grace is a powerful example of how the environment that Liz and David created for her allowed her to build a stronger skeleton.

And if she can surmount her genetic destiny, so can we. Because, though you probably don't know it, just like Grace's, your bones are also breaking all the time. A little crack here, a little fissure there—our bones are in a constant state of deconstruction and reconstruction. In this way, we're all growing more perfect skeletons.

TO UNDERSTAND how DNA is involved in making and breaking our bones, we first need to understand how our bones work. Far from being composed of the dense, dead, and rocklike material many people imagine when they think of bone, our skeletons are quite alive—and are being constantly redeveloped to meet the changing demands of our lives. This remodeling and reshaping comes as the result of a microscopic battle between two types of cells, osteoclasts and osteoblasts, that resembles the relationship between two key characters in Disney's videogame-inspired movie *Wreck-It Ralph*.

Osteoclasts are the Wreck-It Ralphs of the skeletal body, breaking down and dissolving bone piece by piece because they've been programmed to do so. Osteoblasts are the Fix-It Felixes—they have the onerous task of putting your bones back together again. Now, you might think that simply removing Ralph from the equation would result in stronger bones. But that's not how it works. As the characters found out in that charming movie, one can't well exist without the other.

The wreck-it/fix-it partnership results in a complete renewal of our skeletal structure every decade or so. Like a bladesmith folding layer after layer of steel to forge a resilient sword, the break-and-repair-break-and-repeat cycle of bone regeneration leaves us with utterly personalized skeletons that can, in most cases, withstand a lifetime of running, jumping, hiking, biking, twisting, and dancing.

Of course, a little added dietary calcium is usually helpful. And if you're like many people who love breakfast cereal, you get a helping of that almost every morning.

If you eat Froot Loops, Frosted Flakes, or Rice Krispies, you're familiar with the products made by the company founded by William K. Kellogg, brother of the better-known Dr. John Harvey Kellogg. But Dr. Kellogg did a lot more than lend his name to a brand. In his time, he was known as a health guru, though today we'd probably call him a little eccentric. (Among other things, he believed that sex, even the monogamous sort, was dangerous.)

He was also a pioneer in the field of whole body vibration therapy. In his notorious sanatorium, Kellogg subjected his patients to vibrating chairs and stools in the hopes of improving their health. More or less, Kellogg's idea was that he could shake the sickness out of his patients.

More than a hundred years later, vibration therapy is still often viewed with skepticism. Some medical experts have specifically warned against long-term exposure to vibration for most people. But for specific patient groups, researchers are now exploring the possibility that vibrations might trigger osteoclasts and osteoblasts to work together to break down and repair bone. Which is why a therapy long ago rejected as outlandish is now being investigated for use with patients with OI. That, in turn, has prompted another look at vibration therapy for patients suffering from osteoporosis—something that impacts millions of people—by triggering the right genetic expression to make stronger bones.

EVEN FOR those who have a perfect genetic inheritance, disuse, old age, poor diets, and hormonal changes can all wreak havoc on the exquisite balance that shapes our hidden structure. What we're learning is that our skeletal system can be quite unforgiving toward our behavioral indiscretions.

As we're discovering, so too can genetic mutations. Take young Ali McKean, for instance. She suffers from a rare genetic condition that turns her endothelial cells (those that line the interior surface of blood vessels) into osteoblasts (the Fix-It Felix bone-production cells). In other words, her cells are turning her muscles into bones. And yes, that's as terrible as it sounds.

The most famous case of *fibrodysplasia ossificans progressiva*, or FOP, which is sometimes known as stone man syndrome, involved a Philadelphian named Harry Eastlack, whose body began to stiffen when he was five years old and, by the time he died at age 39, was fused so completely that he could do nothing more than move his lips. Today you can visit Eastlack's skeleton at the Mutter Museum of the College of Physicians of Philadelphia, where it continues to interest researchers trying to unlock the mystery of FOP.

Stone man syndrome is thought to affect about one in two million people, and it is aggravated by injury. That means that whenever Ali gets a bump or a bruise, her body responds by sending osteoblasts to the scene of the injury to make bone—and surgery intended to remove the excess tissue causes even more bone to grow back.

In the past few years, those studying FOP have become energized by the discovery that mutations in a gene called *ACVR1* can cause FOP.[4] Some of these mutations are thought to result in a protein switch being made from the *ACVR1* gene, which is always turned on. Instead of having healthy bone growth when and where it's normally needed, it can throw the process of bone growth into overdrive.

As of yet, though, the discovery of the gene is just the beginning in a long road to a cure for those who suffer from Ali's condition. Early detection is key, as it provides a notice for parents and caregivers to help sufferers avoid injury as much as possible. Unfortunately, doctors didn't know what was wrong with Ali until she was five—and if you think about all the bumps and bruises young children suffer, you can imagine how devastating that delay will be for her long-term health. That's not to mention all of the medical procedures she underwent as her doctors struggled to understand what was going on within her, unknowingly doing far more harm than good.

Most of the mutations in the *ACVR1* gene are thought to be new, which we call *de novo* and therefore not inherited from either parent. This only complicates and delays the diagnostic process further since there would likely be no family history of anyone having FOP.

And yet, sadly, there was a clue, albeit a subtle one that was understandably missed: Ali's big toe, which was very short and bent toward the others.[5] That dysmorphic sign, coupled with Ali's other symptoms, could have been understood as a warning that might have helped clinch the diagnosis.[6]

Think about that: Confronted with an amazingly complicated genetic disease, the least invasive and least technologically sophisticated approach to the problem—a long, hard look at Ali's big toe—might have been the best approach to diagnose her condition.

EVEN LONG after we're gone, our bones can leave behind clues about the myriad experiences of our lives that have been impacted by our genes. Harry Eastlack's well-studied skeleton is one obvious illustration—Mutter Museum visitors can see very clearly the way his disease fused his skeleton like a spider wrapping a fly in its web. But there are other, far more subtle examples.

For instance, let's say we had recovered some bones from the long-lost crew of the *Mary Rose*, the sixteenth-century English naval flagship of King Henry VIII, which sank on July 19, 1545, while fighting a French invasion fleet. What would those bones tell us?

Although there are a lot of differing accounts, we still aren't sure why the *Mary Rose* sank, nor do we know much about the identity of the men whose bodies settled to the bottom of the Solent Strait, just north of the Isle of Wight in the English Channel. But a modern scientific process called osteological analysis can help us decipher how their bones were used. And the *Mary Rose*'s sailors left us one huge hint: They had large left shoulder bones.[7]

Researchers believe most of the physical tasks demanded of the sailors would have had them use both hands equally. Except, that is, for one important task—longbow archery was mandatory for all able seamen in Tudor England, and the *Mary Rose* carried 250 bows onboard (many of which, it appears, were used to shoot "fire arrows" at enemy ships).

Unlike today's carbon competition-caliber bows—the complex mechanical types you might see in the Olympics—the ones that were used in sixteenth-century England were very heavy. And while many things have changed in the centuries since the *Mary Rose* sank, one thing hasn't. If you're right-handed, as most of us are, you're more likely to carry your bow with your left hand.[8]

Of course, we already know that the repeated use of one arm over the other can result in differences in muscle shape, size, and tone. If you play tennis or just watch it closely, you know that a player's racket-wielding arm is often significantly more muscular than the opposite arm. (Left-handed Spanish phenomenon Rafael Nadal is a great example of this—his dominant arm looks like it belongs to a smaller and less green version of the Incredible Hulk.)

But constant use, strain, and weight isn't just toning muscle, it's also setting osteoclasts and osteoblasts to work, which changes genetic expression that helps to build stronger bones. It's also weaving together an aspect of our life story that will last as long as our bones do.

We don't have to look back hundreds of years to see an example of our malleable skeletons at work. If you've ever laid eyes on a bunion you've witnessed the effects of the same phenomenon. Sitting on the Metropolitan Transportation Authority's number 6 subway line during its run through Manhattan in the middle of summer when everyone's wearing sandals provides one of the best opportunities for bunion viewing. If you've got one, or ever get one, don't curse your bones for misbehaving—they're only responding to the life of constrained foot apparel they've been subjected to. Not to mention an unfortunate genetic predisposition that seems to prime you for them.[9] So don't beat yourself up if you end up with bunions. Instead, this might be the only time when you can rightfully simultaneously blame both your parents and your fashionable shoes.

As we've seen, regardless of our genetic predisposition, for the most part we've all inherited genes that allow us to have malleable skeletons. Another example of how our behaviors can cause changes to our bones is at play in the lives of our children. For years now we've begun to notice detrimental changes in the curvatures of the spines of elementary-age schoolkids, who have been paying the price for overloaded backpacks.[10] As a result of increased attention to this problem, many parents have given their children packs with wheels, not unlike the carry-on suitcases many of us take with us to the airport.

Not surprisingly, a lot of kids have bucked at taking rolling bags to school. "Dorky" is how my friend's middle-school-aged son put it. That's why one company's inventive response to the problem—a scooter that folds, Transformer-style, into a backpack with wheels—has been a gold mine. Two years after launching its product online, Glyde Gear was still so flooded with demand that it was taking more than a month and a half to fulfill old orders and had to temporarily stop taking new ones.

Not all good intentions go without consequences, though. Traditional backpacks were bad for kids' posture. Roller bags, it seems, present a tripping hazard and school maintenance headache (they tend to scuff up floors and ding walls).

Unfortunately, that's often what happens when it comes to medicine, too. As we will see in the next few pages, new solutions to old problems often create new problems in need of even newer solutions. And sometimes by being too flexible, as when our bones are too malleable during our early years of development causes them to become permanently misshapen.

AN EXAMPLE of this started happening in the mid-2000s in response to the National Institute of Child Health and Human Development's Back to Sleep campaign. Thanks to this successful initiative, the percent of parents dutifully putting their babies to sleep on their backs has soared from 10 percent only a few years earlier to a whopping 70 percent today.

The campaign was born in response to recommendations from the American Academy of Pediatrics, which had been seeking to reduce sudden infant death syndrome (SIDS) cases by changing habits associated with a problem that was claiming the lives of about one in every 1,000 infants.

Over a 10-year period following the introduction of the campaign, the rate of SIDS death fell by half. As with any medical innovation, with that success came a rather unforeseen but thankfully somewhat benign complication. Babies who sleep on their backs, while the boney plates that form the back of their skulls are still forming and fusing, become more likely to have slightly misshapen heads. And babies with misshapen heads became far from exceptional: During the years in which back sleeping became the norm, the incidence of such affects quintupled.[11]

The technical term for this benign phenomenon is *positional plagiocephaly*, and for the most part we don't consider it to be a very big deal medically. But with our society's increased obsession with physical perfection, many parents have resorted to visiting an orthotist, a specialist in external devices designed to modify the functional or structural characteristics of our bones and muscles. Using something called a cranial remodeling helmet, orthotists can help correct a baby's head shape. Positional plagiocephaly is an example of how our bodies are not functioning in a developmental vacuum but can be induced to permanently change in response to the circumstances of our lives.

My first encounter with such a helmet came about a decade ago while I was walking through Central Park in Manhattan. At the time I had no idea what they were for and wrongly assumed that I was witnessing a new fad among very safety-conscious parents of children being helmeted while in their strollers.

I eventually learned the details of how it works. The purpose of the helmet is to reshape a kid's skull by removing pressure over the flatter parts, allowing the skull to grow into those areas. This device works best for children between four to eight months of age, needs to be worn 23 hours a day, and must be adjusted every two weeks. They can cost upward of $2,000 and generally aren't covered by insurance.

Because their children's heads are quite malleable, though, studies are showing that parents who use stretching exercises and special pillows can see significant improvements in their child's head shape even without the use of a helmet.[12]

In the long run, though, the important thing isn't shape but strength. As a species, we're a rather clumsy lot—and given the importance and relative fragility of our brains, it's vital that our skulls retain structural integrity.

But strength isn't just a matter of material hardness. When it comes to our bones and our genome—real strength lies in flexibility. Which is why I want to tell you about Michelangelo's *David*.

IT WAS like walking into a photo by Edward Burtynsky.

The much-lauded photographer, famous for his images of industrial landscapes, has spent a lot of time taking pictures of Italy's Carrara marble quarries, which are renowned for the beautiful and plentiful white-blue marble that is harvested there and used by builders and sculptors around the world.

As I traveled through the Italian Alps a few years ago and came upon one such quarry, I marveled at the audacity of the operation. Enormous tractors crawled along narrow mountain roads, moving marble blocks the size of minivans from deep inside the earth to preparation centers in nearby Tuscany. From there they travel by train, ship, and truck to many points across the globe.

Marble is a product of the metamorphosis of sedimentary carbonate rocks formed millions of years ago as seashells settled on the bottom of the ocean. These sediments eventually became limestone, and after millennia upon millennia of the heat and pressure of tectonic processes, it is finally freed by operations like those in Carrara.

Carrara marble is a relatively soft rock and easy under the chisel, which is why it is so sought after by sculptors and artisans. It is also very strong, which is why sculptures like Michelangelo's *David* have survived intact for more than 500 years.

Well, mostly. As it turns out, David has bad ankles, and over the years, the pitter-patter of millions of tourists' feet at Florence's Galleria dell'Accademia have taken their toll on the statue's stability. In a way, David's strength is his weakness—the inflexibility of his marble leaves him vulnerable to cracking.

That is how we'd be, too, if not for our regenerative skeletons and the genes that code for things like collagen that gives them their structure.

In humans, the production of collagen depends on our DNA and is produced in response to the demands imposed by our lives. Unlike Michelangelo's *David*, our own ankles can heal after a sprain thanks to an increase in collagen being made through genetic expression.

In humans, collagen comes in more than two dozen types, and besides being essential for healthy bones, is found in everything from cartilage to hair to teeth. Of the five main types, type I is the most abundant; it makes up more than 90 percent of the body's collagen. This type of collagen is also found in artery walls, giving them the elasticity necessary to keep them from bursting every time our hearts contract and kick out a ventricle's worth of blood.

If there's one place that we all really notice when collagen begins to fail and lose its tensile strength, though, it's in our faces, where it provides structure to our skin. That's why, when you hear of collagen, you might think of it as the substance that some people have injected into their cheeks to make them look younger.

And that's not a bad place to start, because it helps us understand the role that collagen plays as a structurally supportive protein. After all, no one would use it to create puffier cheeks and fuller lips if it wasn't going to hold the shape, right?

The word collagen originates from the ancient Greek word for glue, *kolla*. Before the modern industrial production of glue, most people had to rely on their own know-how to keep things bonded together. And it was glue made from the boiling of animal sinews and skins (rich in collagen) that was the source of strength in the bonding process. (This is where the expression "sending the horse to the glue factory" comes from.)

Catgut, which is used in making strings for classical musical instruments, is also made mostly of collagen found in the walls of the intestines of goats, sheep, and cattle (but not, as it turns out, from cats). It has also been used, for many years, to make tennis rackets; it takes around three cows to make the strings for just one racket. It's the tensile strength, derived from collagen in the serosa of animal guts, that makes catgut so desirable. Tensile strength is the measurable force at which a material can be stretched or deformed before failing. It can be thought of as the opposite of a substance being brittle.

It's also what makes certain foods so much fun to chew. If you're into sausage or like to grill hot dogs for summer barbecues and tailgate parties, you'll be happy to know that all the various parts and pieces used in the making of many frankfurters are held together by the super strength of collagen. And as many vegans will tell you, Jell-O, marshmallows, and candy corn all get their texture from gelatin, which is also derived from collagen. All told, some 800 million pounds of gelatin is produced every year worldwide and makes its way into your home or palate through different routes, from frosted Pop-Tarts to vitamin capsules and even certain brands of apple juice.

From striking a ball with a tennis racket, to pinching the cheeks of a loved one, to the bouncing here, there, and everywhere joy of gummy bears, that elastic "snap back into shape" action you're feeling is all thanks to collagen.

The ultimate example of how flexibility equates to strength, though, is a two-meter-long freshwater fish called the arapaima. It is among the few animals that can live without fear in piranha-infested waters, thanks to genes that encode for collagen-backed scales that give, but don't break, when struck with sharp objects. Researchers at the University of California at San Diego figure this makes the arapaima—which hasn't evolved much in the past 13 million years[13]—a good model for building flexible ceramics that can be used in body armor—just one of the many ways that turning to the natural world for solutions can help us solve problems relative to our modern lives.[14]

HOW DOES all this relate to genetics? Without our genome's inherent flexibility, our bones become ill suited for the rough-and-tumble lives we lead. And as we've learned with Grace, Ali, and Harry, it doesn't take much to throw everything out of whack.

In fact, all it really takes is a single letter.

The human genetic code is made up of billions of nucleotides—adenosine, thymine, cytosine, and guanine, which we abbreviate with the letters A, T, C, and G—all lined up in a very specific pattern.

Now, within the area that normally codes for building collagen in our bodies, in a corresponding gene known as *COL1A1*,[15] the code generally goes a little something like this:

G A A T C C—C C T—G G T

But a single random mutation can make it look like this:

G A A T C C—C C T—**T** G T

And that's all it takes for our body to change the way it makes collagen. One letter off in the code, and instead of a strong, and flexible skeleton, we get bones as stiff as marble, or brittle as sandstone.

How could one single letter make such a profound difference?* Well, imagine for a moment listening to Beethoven's well-known piano composition "Für Elise." It begins as it always begins. But when the pianist gets to the tenth note, she misfires. Not by a lot, just by a little. Would you notice? Would the piece be the same? And if you were a classical music producer, recording the rendition for posterity, would you simply ignore that mistake?

---

* In the example given, the single nucleotide change turned out to be deadly, causing a lethal form of osteogenesis imperfecta.

Beethoven was brilliant. His compositions were incredibly intricate. But compared to your genetic code, even Beethoven's greatest masterpieces were as complicated as "Mary Had a Little Lamb."

Our code is like a journey of billions upon billions of steps. If the first one is just slightly askew, the rest of the journey will be, too.

And so, in a very real sense, we are all one letter away from having a life-altering genetic condition. But just as we saw with Grace, that doesn't mean we're completely helpless, either. As we'll see in a lot more detail, getting off the couch does a lot more than just get your body moving.

WHAT WE don't use, we lose. And rather quickly, too.

Just as the most efficient businesses have employed just in time strategies that match industrial production to demand in near real time, our species has evolved genetically to keep the cost of living down, reducing inventory when we don't need it and hyperproducing it when we do.

That's one possible reason why older obese people are less likely than their thinner peers to suffer many kinds of common fractures. They resemble ancient archers carrying around extra weight. The added wear and tear on their skeletons throws their osteoclasts and osteoblasts into a furious break-it-fix-it cycle that can result in stronger bones.

By way of contrast, we also know that swimmers, whose athletic endeavors take place in reduced gravity environments, have lower femoral-neck bone mineral density than athletes who engage in weight-bearing activities,[16] likely because swimmers (though they get an incredibly beneficial cardiovascular workout) just don't take the same kind of skeletal pounding as athletes in other environments, such as runners and weight lifters.

We get another example of this whenever space travelers return from the International Space Station after long voyages. When a Soyuz space capsule carrying U.S. astronaut Don Pettit, Russian Oleg Kononenko, and Dutchman André Kuipers landed in southern Kazakhstan in July of 2012, ending a six-month sojourn in space, the three men had to be gently hoisted into special recliners for post-mission press photographs.[17] During 193 days of swimming through the weightlessness of space, their bodies had begun to chip away at their skeletal solidity.

In this way, astronauts are a lot like osteoporotic old women. And, as it turns out, their medical care is a bit similar. Bisphosphonates such as zoledronate and alendronate (drugs that essentially convince osteoclasts to kill themselves instead of breaking down our bones) are a mainstay of treatment for older individuals with osteoporosis. And we've recently learned that these same drugs can also help both astronauts and people with OI keep their bones in better shape.[18] With the news that some private companies are looking for volunteers to make the first human mission to Mars—a journey that will likely take a minimum of 17 months in a zero-gravity environment—those drugs will be vital.

But before you volunteer to step onto that spacecraft, a little warning. Although people who are taking bisphosphonates do become less susceptible to fractures where they usually happen in the elderly, at the femoral neck, they conversely become more susceptible to fractures in the shaft of the bone.

Why? Because the drugs are actually working too well—stopping bone turnover and remodeling, leaving users with something called "frozen bone," which is thought to increase susceptibility to certain types of fractures, just like the statue David's ankles.

I'M ALWAYS left awestruck by the incredible range of effects that emanate from the subtlest of changes in our genetic code and its expression. As we've now seen, one single letter change in a series of billions of letters, and you've got bones that break with the slightest bit of pressure. A small shift in any of our genes can completely change the course of our life.

And if you have inherited a faulty gene, or stay in bed long enough, don't exercise, eat poorly, escape gravity, or simply get older, you'll be setting yourself up for similar detrimental skeletal consequences. With a growing list of options that include an arsenal of medications, weight-bearing exercises, and maybe even vibrational therapy, we are far from helpless skeletal custodians. Whether the vulnerability is related to genes, lifestyle, or both, there are many preventative and therapeutic modalities available that we can employ to make our bones less susceptible to fractures.

Understanding the basic biology of how we lose our bones can also play an important role in shaping how we learn to keep them. Such knowledge can be used to inform our own life choices, guiding us onward to pursue activities and lifestyles that will build us the strongest of skeletons.

To do that we need to discover the entire genetic underpinnings of how our bones work. By studying Grace and other people whose DNA leads to brittle bones, we can arrive much more quickly at newer treatments for far more common conditions such as osteoporosis.

When it comes to genetics, the rare informs the common.

And in so doing, millions of unsung heroes like Grace are gifting the entire world with an incredibly precious genetic gift.

CHAPTER 5

Feed Your Genes

What Our Ancestors, Vegans, and Our Microbiomes Teach Us about Nutrition

I fell asleep in my day clothes. Sometimes, after a particularly long shift at the hospital, that's just the way it goes. It takes every bit of energy I have just to get home, walk up the stairs, and fall into my bed—and pajamas feel like a luxury I simply can't afford.

It was a little past midnight when I crashed atop the covers. I could have sworn that only a few minutes had gone by when my pager began buzzing against the bedside table.

With my face still buried into my pillow, I reached for the cursed little black box. When I couldn't immediately find it, I reluctantly craned my neck and opened my eyes. The glowing blue numerals on my alarm clock flickered from 3:36 to 3:37 a.m.

Three and a half hours, I thought to myself, already trying to calculate how much wakefulness I'd secured with a half-night's deposit of sleep. *That's not so bad.*

It doesn't take many wee-hour pages before you begin to recognize the numbers: 175075 is the emergency department, 177368 is the inpatient ward. And 0000 means there's an outside caller, on hold, waiting to speak with you.

The challenge with calls like that is you never know what to expect. Sometimes it's concerned parents who already know their child suffers from a rare genetic disorder but who aren't sure whether a new set of symptoms they are seeing is something to be worried about. Other times it's a physician from another hospital who has just seen a patient they're having trouble figuring out how to treat, and so it's a call for advice. In some cases the call is one no physician wishes to get at any time—a patient has taken a turn for the worse.

I grabbed my phone and tried to sneak out of bed without waking my wife, who was sleeping softly beside me. Tiptoeing out of the bedroom, I gently closed the door, peeking back through the crack as I exited. No mumbles. No restless thrashing. She was still out.

Success! I am a nighttime ninja.

I pressed the recall button on the pager. The dreaded 0000 stared back at me like two tiny sets of owl eyes. The bright blue numbers lit up the dark hallway. I dialed the number and waited.

"Hospital locating…"

"It's Dr. Moalem. I'm on call for—"

"Thank you for returning the page. Connecting…"

There was a soft beep, then a torrent of words.

"Dr. Moalem? I'm sorry, I know it's late…or is it early? Either way, I'm sorry to bother you. It's just that—my daughter Cindy. She's had a fever the last few hours and I'm worried because she hasn't been eating very much today."

To some people, this might sound like an overanxious parent. But I knew the hospital wouldn't have patched her through to me if there hadn't been more to it.

She paused for a moment. Instead of breaking in, I let the silence hang over the line.

"Oh—I should have mentioned," the woman said. "My daughter has OTC deficiency."

There it was. In ornithine transcarbamylase deficiency, or OTC, a rare genetic condition affecting about one in every 80,000 people, the body struggles to work its way through the process of turning ammonia* into urea, which under normal circumstances is quickly expelled from our bodies when we urinate.

This process, called the urea cycle, takes place mostly in the liver and, to a lesser extent, in the kidneys—it's a barometer of sorts for our overall metabolic health. When it's working right, we're doing what needs to be done to metabolize protein. When it's not working right, our bodies become loaded with ammonia—which is every bit as nasty as it sounds.

And like a factory pumping out toxic waste, the greater the metabolic demand, the greater the level of waste ammonia produced. Which is exactly what normally happens when we have a fever. For every two extra degrees Fahrenheit of increased body temperature, our systems burn through about 20 percent more calories than normal. Most of us can handle that extra demand for a time. In fact, for most people a little fever when ill does some good, raising the body's temperature just enough to make life quite difficult for some illness-causing microbes, thus slowing their growth and giving the body a chance to fight back.

But for people like Cindy, whose systems are more precariously balanced to begin with, just a slight low-grade fever and things can go very badly, very quickly. The nervous system is, after all, quite sensitive to rising levels of ammonia and falling levels of glucose, which we use for energy. And if left unchecked, this metabolic situation can cause seizures and organ failure, which in turn can lead to coma.

In other words, Cindy's mother had good reason to be concerned about her daughter. And I had good reason to get out of bed.

---

* A common by-product of the metabolic processes that occur when the body breaks down protein.

I grabbed my laptop and tapped in the pass code that lets me log in remotely to the hospital's system. Given Cindy's previous history—over the past few years she had been hospitalized multiple times—it was clear she needed to come to the emergency department.

Thankfully, her family lived nearby.

I'm close, too—doctors on call who choose to live more than a few minutes from the hospital often come to regret that decision. I packed a few things into a knapsack and counted myself fortunate that I didn't have to sneak back into my bedroom to change my clothes—because the truth is that I'm not actually a ninja. When it's dark, I'm pretty clumsy—and noisy. In the wee hours of that morning, at least, my wife could stay warm, snug, and undisturbed in our bed.

I grabbed a banana from the kitchen counter and headed out the door. It wasn't yet 4 a.m., but I was wide awake.

AS I drove to the hospital and snacked on my banana, I considered how privileged I was not to have to fret too much about the food I eat. Like most people, I try to keep my sugar and fat intake down. On rare occasions, when I'm feeling gastronomically bold and mathematically capable, I try to balance a breakfast, lunch, and dinner that will help me hit 100 percent on all 21 vitamins and minerals recommended by the Food and Nutrition Board. Try it sometime—it's harder than it seems.

Truth is, though, that a diet based on those recommendations alone is hardly perfect for most people. In fact, it's extremely unlikely (as in you've got a better chance at winning the lottery) that the portion recommendations and percentages you've been studying on the sides of prepackaged foods are anywhere close to exactly right for your individual needs. That's because those numbers are based on a sweeping estimation of the necessary intake of calories, vitamins, and essential minerals for a majority of the healthy people in the United States over the age of four. (And for the Food and Nutrition Board, "majority" means 50 percent plus one person; which leaves a tremendously big minority for whom the guidelines are simply a nonstarter.)

The reality, of course, is that everyone's needs are quite different. The majority of four-year-old boys (for whom 275 micrograms of vitamin A each day will generally suffice) are very different from the majority of 32-year-old pregnant women (who generally need at least three times as much vitamin A). Even two people of the same sex, age, ethnicity, height, weight, and general health are likely to have very different needs when it comes to calcium, iron, folate, and a host of other nutrients. The study of the ways in which our genetic inheritance impacts our dietary needs is called nutrigenomics.

In chapter 1, you met Jeff the Chef, who suffers from hereditary fructose intolerance (HFI). That's a relatively rare disease, but to some extent all of us could benefit from getting to know the genes within our genomes. And for the millions of people who have some type of unique nutritional requirement that is impacted by their genes, it's not rare at all to feel as though food isn't your friend. That is why there are a lot of people out there with similar conditions that make minefields out of restaurant menus and gauntlets out of grocery lists.

Now, you might recall that HFI requires people like Jeff to craft personal menus devoid of fruits and veggies (and also fructose, sucrose, and sorbitol, which are often added to processed foods). Cindy's OTC deficiency is a sort of dietary contrapositive to that. People who are mildly affected by OTC can often go undiagnosed.

They will frequently say they just don't feel good when they eat meat, and they've thus avoided heavy protein meals their whole lives. Genetically speaking, they're actually much better off as vegetarians or vegans because it can be easier to manage a lower protein intake.

Not unlike our political beliefs, which can run the gamut from anarchism to totalitarianism but generally fall somewhere in between, our diets exist on a wide and diverse spectrum. Just as most of us can tolerate many political ideas that we mildly disagree with, our bodies are generally good at stomaching most types of food. And just as there are some ideas that you probably cannot abide—the repeal of universal suffrage, for instance—there may be a few foods that are simply incompatible with your genetic makeup.

Many of us may not have spent a great deal of time reflecting on the inner workings of our political views, let alone examining how we adopted those beliefs. In the same way, chances are good that there are foods that your body just doesn't like—and chances are also good that you don't know why.

That's starting to change, though. In recent years, people who are concerned that they might have health problems tied to the foods they eat have found initial help in elimination diets in which they reduce their food intake to a small number of foods and build slowly back from there. The educational equivalent might be an introduction to political philosophy class that exposes students to the evaluation and history of a wide array of social and governmental ideas.

There's just one problem: The solution is not that simple.

FOR NOW, a lot of us have simply resigned ourselves to eating the way our doctors have long told us to: Eat lots of this and none of that; eat this occasionally and that rarely. And for most people that advice is at least a good starting point.

Just as our politics are often a reflection of our regional and cultural heritage, so too originally our diets were a reflection of our genetic inheritance.*

For most people of Asian descent, for example, milk and dairy products aren't just an unpalatable proposition—they can be digestively hostile. You see, if your ancestors kept animals for milk** there's a high likelihood that their genes sustained mutations that now make your genes excellent manufacturers of the enzymes necessary to break down lactose, one of the sugars naturally found in milk, long into your adulthood. But in most other parts of the world, where dairy livestock are not as historically common, lactose intolerance in adulthood is much more prevalent.

In spite of this, China has seen a tremendous rise in dairy product consumption over the past decade. Not surprisingly, though, the Chinese tend to prefer types of hard cheeses, or local varieties such as *rubing* (a delicious goat milk cheese from Yunnan Province similar to the Mediterranean halloumi). That's because, unlike softer cheeses like ricotta, hard cheeses are generally easier on the lactose.[1]

In a way, eating how your recent ancestors did is akin to the way today's patients' medical family histories are useful tools to assess patients' current health risks. If you come from an ethnically diverse heritage and use this approach to assess your dietary needs you can end up with some pretty interesting genetic and culinary fusions. Which can sometimes lead to confusion and frustration, especially given the ethnic genetic melting pots so many of us now come from. For ex-

---

* Even if you do know the types of foods your recent ancestors ate, you need to take into account that they might be too calorically rich (I'm thinking lard in apple pies, for example) for today's comparatively lower levels of physical activity.

** And if you're of West African or European descent there's a good chance they did.

ample, many Hispanic individuals are a mix of a myriad of genetic fabrics. If you're Hispanic, whether you're lactose intolerant depends upon which part of the genetic ancestral quilt you've inherited.

Then again, regardless of whether we come from one ethnic and cultural background or 16, these days almost all of us have palates that have become somewhat globally oriented, which has the potential to overtake what we may actually nutritionally need. In the developed world, even the smallest of grocery stores in the sleepiest of towns offers a selection of meats, fruits, and grains that our not-too-distant ancestors, even if they were royalty, couldn't have dreamed of accessing.

Following my own advice and looking for dietary guidance from my recent ancestors means heartily consuming a bowl of semolina gnocchi stuffed with walnuts and dates, and knowing that all will go well digestively as a result. Of course, your personal definition of palate exploration may look quite different. And if you haven't tried recently changing what you eat, now may be a great time to grab a plate and take a seat at your ancestors' table. Given our comparatively sedentary lifestyle today, though, we're going to have to use a much smaller plate as well.

Even with persistent dietary experimentation, we're still going to have to contend with the fact that changing attitudes and habits about food is a lot of work. To help you on that journey it's useful to know that some studies have found that when we combine theoretical education with experiential "cook and eat" teaching sessions (not only bringing the horse to water but showing it that the water can taste refreshingly good, too), the chance for successful integration becomes better.[2]

And, of course, there's another significant motivator—the same one that got former President Bill Clinton inspired to change his diet a few years back: the ubiquitous desire to live a long, full, and healthy life.

For the former president, after a lifetime of eating whatever seemed good at the time, undergoing two heart surgeries, and taking stock of a family history of heart disease, Clinton finally decided to make some serious life changes, including moving to an almost completely vegan diet, in 2010.[3] Sometimes you just have to be pushed into making a total change and, just like Clinton, radically change your nutritional lifestyle. Even if you're properly motivated, accessibility and the affordability of nutritious and healthy foods can come with substantial obstacles, but ones that are worth working to overcome.

Okay, what have we learned so far? Find good food, eat like your recent ancestors—but not as much—get physically active, and then listen to your body for telltale clues that you're on the right path.

If only life were so simple. Far from being a utopian solution, eating strictly like your ancestors will not work for everyone. We are, after all, genetically unique. In fact, as we saw with Jeff the Chef and Cindy who has OTC deficiency, not taking stock of what we individually inherited can even turn deadly. Each of us should be eating in a way that more closely matches our distinctive genetic inheritance.

As we're about to discover, this is far from a modern problem—something that our seafaring ancestors would have no trouble recognizing, either.

ENSHRINED IN nutritional lore is the story of how British sailors suffered horribly from bleeding gums and easy bruising—a condition known as scurvy—because of the lack of fresh fruits and vegetables aboard maritime ships. Back before we'd figured out electrical refrigeration, the best sailors could hope for was a combination of cured and dried meats and some crusty bread. For men stuck at sea for months at a time, this led to some pretty nasty nutritional deficiencies—and, curiously, not all sailors suffered equally from them.

Today, we know that citrus fruits are rich in vitamin C, which for most people is good for preventing the sorts of deficiencies that some of those sailors faced. Back then, they just knew that lemons and limes could keep their teeth in their mouth and the other symptoms of scurvy at bay.

Interestingly, the rats on those ships didn't have that same problem. Neither did the cats that were often kept on board to do battle with the rodent marines. So why were the rats and cats not losing their teeth as well?

From aardvarks to zebras, most of our mammalian cousins have working copies of the genes that can manufacture vitamin C naturally within their bodies. But humans (along with guinea pigs, of all things) have a genetic inborn error in metabolism, a mutation that renders us incapable of doing the same thing. This makes us completely dependent upon our diets to get our daily supply of vitamin C.

Some small groups of seafarers appear to have figured out the magic of citrus fruits centuries earlier, but it wasn't until the end of the eighteenth century that the British Admiralty, encouraged by a Scottish physician named Gilbert Blane, had its sailors drink lemon juice to combat scurvy. And on return trips from the empire's Caribbean territories, where limes were more plentiful, the ships were loaded with the lemon's green taxonomic cousin—which is how British sailors became known as limeys.[4]

Once we figured that out, though, it was only natural to want to determine the minimum amount of lemons, limes, oranges, and the like we need each day to stay healthy (after all, the famously bureaucratic Brits needed to know just how many citrus fruits to pack for a long voyage at sea). This is the root of modern nutritional science, which to this day is based on the idea that we can math our way to a healthy diet. Hence the "reference daily intake" (formerly known as the recommended daily allowance) that is used to determine—down to the gram, milligram, and microgram—the daily values of food we all supposedly need to live healthy, active lives. Many of these values have been derived from what's needed for the average person to overcome a symptomatic deficiency and not what's optimal for us as utterly unique individuals.

Which is why we don't all need the same amount of vitamin C. As we move toward what's individually optimal, we'll have no choice but to look at our genes. In a study that focused on the genes that help get vitamin C into our bodies, researchers found that variations in a transporter gene called *SLC23A1* affected the level of vitamin C completely independent of our diet.[5] And even with a corresponding higher intake of vitamin C, some people, it seems, will always have lower levels of vitamin C no matter how many citrus fruits they eat. Discovering which version of the transporter gene we inherited can have a tremendous impact on our understanding of the amount of vitamin C that gets successfully absorbed into our bodies.

Direct dietary advice is not, however, all we need. We are discovering that some of the differences in our genetic inheritance, for instance versions of another gene involved in vitamin C metabolism, *SLC23A2*, have been associated with an almost threefold risk of spontaneous preterm birth.[6] It's been suggested that this might be related to vitamin C's role in the production of collagen, which helps provide the tensile strength a mother needs to keep her baby inside her body,[7] again underscoring the importance of taking our genetic inheritance seriously when it comes to nutrition.

So, given that generalized dietary advice can be wrong when it comes to the individual, you may rightly be wondering how much is the right amount of citrus. And what's the right diet for you? And what foods should you avoid? The answers to those questions are going to be different for everyone, not only because of the genes you've inherited but because, more importantly, what you eat can completely change how your genes behave.

THIS YEAR, tens of millions of Americans will attempt to change their diet.

And most of them will fail.

In part, that's because without knowing what diet is genetically right for them, some of those folks are essentially flying blind—and many are doing things that are counterproductive to their goals.[8]

But even for those in the majority, for whom the advice to eat a reasonable diet and get vigorous exercise is still the best medicine, there's another problem: Dieting is hard.

For most of human history, food was far from plentiful. To mitigate this scarcity, coupled with the rare times that food became bountiful, we've all inherited genes that favor overeating. And in the past, if there happened to be any caloric leftovers from those rare big meals, our bodies would eagerly pack them away as body fat. Like a caloric savings account, storing what we didn't use came in handy when scarcity returned. And for most of human history, we've known more paucity than plenty.

Today we're faced with a complex problem, a glaring mismatch between what we've inherited and the current environment we find ourselves in. First off, given our sedentary lifestyles, we don't need anywhere near the same amount of calories to get by as in the past. We've sentenced machines to do most of our hard labor for us and get us around from place to place. And secondly, combine that with the abundance of cheap available calories, and it's easy to understand why obesity rates are soaring today as never before in human history.

It's not just the amount of foods we're consuming, either. As we'll come to see, our food choices are far from optimized for our genetic inheritance.

Thanks to the science known as nutrigenomics, we're starting to figure out what to leave off the individualized contemporary menu. For example, you'll no longer need to wait to get bloated, write a food diary, and have diarrhea to find out that you are lactose intolerant. The genetic test that would give you that information is already commercially available. And if you're an early adopter you might have already looked beyond single gene testing, say for lactose intolerance, and decided to go all the way and get your exome or even your entire genome sequenced.

Which can then be used for twenty-first-century genetically based dietary advice. You can use this information to decide whether your next cappuccino should be caffeinated. This decision would come about by finding out what version of the *CYP1A2* gene you've inherited. Different versions of this gene determine the rate at which your body breaks down caffeine. You may be a fast or slow metabolizer of one of the world's oldest stimulant drugs.

Having a different version of the *CYP1A2* gene and consuming caffeinated coffee can have far more wide-ranging effects than just keeping you up at night. Again depending on the version of *CYP1A2* you've inherited, you're likely to experience a consequential unhealthy spike in your blood pressure. This is thought to happen if you've inherited a copy of the *CYP1A2* gene that breaks down caffeine slowly. On the other hand, if you've inherited two copies of the same gene that burn through caffeine quickly, your blood pressure is not likely to be affected in the same way.[9]

Let's start putting together pieces of what we've learned so far about our genomes and nutrition, because things are about to get a lot more interesting. As we're learning, our lives are not functioning in a genetic or environmental vacuum, with only single-gene interactions. We previously spoke about how our genomes are continually responsive to how we behave and what we eat. Like Toyota and Apple, employing JIT or just-in-time forms of production, our genes are constantly being turned on or off. And this happens through genetic expression—where genes are induced to make more or less of a product.

An example of how our lives can affect our genes in interesting ways can be seen in smokers who drink coffee. Have you ever wondered why people who smoke tobacco seem to have no problem consuming very large quantities of coffee?

The answer has to do with genetic expression.

Our bodies actually use the same *CYP1A2* gene to break down all sorts of poisons. Given its noxious contents, it's no surprise that tobacco is a very loud genetic call to action, and in this regard smoking induces or turns on the *CYP1A2* gene. The more this gene is turned on, the more easily your body can break down caffeine in coffee. Don't get me wrong—I'm not suggesting that you take up smoking so you can drink more coffee and still fall asleep at night. I'm just saying that smoking changes the way your body breaks down caffeine, which can turn a genetically slow metabolizer into a faster one.

Anyway, if coffee doesn't agree with your genetic makeup, you're always welcome to brew yourself some green tea. And before you sit down to enjoy some sencha or matcha, just a quick reminder that nothing we do is without some type of genetic consequence.

In the case of green tea it's been suggested that it may play a role in preventing some forms of cancer. More recently, researchers gave breast cancer cells one of the potent chemicals found in green tea called epigallocatechin-3-gallate, and they noticed two very important results. The breast cancer cells began killing themselves through a cellular process called apoptosis, and those cells that didn't, still showed much slower growth. This is exactly what you want to see happen if you're looking for new treatments for rogue cancerous cells.

When the details were worked out as to how the cancer cells were coaxed to change their behavior, it became clear that epigallocatechin-3-gallate can promote positive epigenetic changes—those on and off modifications to DNA that work to help regulate gene expression. These are important and crucial steps in trying to control cells when they decide to stop obeying our bodies' collectivist biological manifesto. When cells cease to work together cooperatively and go on a malignant rampage, you end up with cancer.

The more we study the interplay between our genes and what we eat, drink, and even smoke, the more apparent it becomes just how important these interactions are to the maintenance of our health.

And from studying monozygotic twins who have inherited the same genomes and are eating similar diets, we're now uncovering the crucial missing piece to our nutritional puzzle.

Which is why it's time that I introduce you to your microbiome.

THE HUMAN gut is a mind-bogglingly complex example of microbial biodiversity.

Two of the main players in this huge little ecosystem are the phyla Bacteroidetes and Firmicutes.[10] If you add up all the species belonging to each of these groups, you'll come up with several hundred different types of microbes—and everyone's microscopic menagerie is a little bit different.

To the microbes living within you, the 30 feet of plumbing from your mouth to your anus is a veritable planet. Its twists and turns, if mimicked in roller-coaster form, would humble even the most experienced thrill seeker. And the difference in conditions, from one part to the next, is like going from the bottom of the sea to the inside of a volcano to the lushest of rain forests.

It's probably not surprising, then, that the gastrointestinal system is one of the most complex structures our bodies construct during fetal development. To give you an idea of the developmental Cirque du Soleil required, at one point in our fetal development, our intestines actually grow out and into the area inhabited by our umbilical cord. To make it back safely into the abdominal cavity, the intestines need to twist and turn, stuffing and coiling themselves like a snake back into a charmer's wicker basket. Which is why it doesn't take much to throw the process off. If the intestines get trapped on their way back into the body, an omphalocele—a sort of intestinal and umbilical herniation—can form. If the intestines do make it safely into the abdomen but the body wall fails to close properly, gastroschisis can occur. This is the name given to the condition that results when parts of the intestines remain on the outside of the body during development, poking out through a crack or crevice. Because the intestines and the amniotic fluid are not meant to meet, the exposed intestines are usually damaged and need to be surgically removed and reconnected.[11] And those are just a few of the many things that can go wrong in the development of a system that, later on, will house a jungle of physiological and microbial flux.

So while it's not always pleasant to think about, it turns out that knowing a bit more about what's happening inside our intestines might be one of the more novel things we can do to stay in touch with our personal health.

To understand this better, let's take a trip to China, where scientists at Shanghai Jiao Tong University recently turned the dietary science world on its ear.

Here's what happened: When studying the gut of one morbidly obese person (who at 385 pounds was about the size of the average sumo wrestler), the scientists noticed an abundance of bacteria that belong to the genus known as *Enterobacter*. Now, lots of people have some *Enterobacter* in them, but in this particular patient, it made up 35 percent of the bacterial forms in his system. That's a lot. So to better understand what was happening, the researchers took a strain of the bacteria from the patient and introduced it to mice that had been raised in a completely germ-free environment.

And, well, nothing happened.

That could have been the end of the experiment. But the Shanghai scientists then decided to see what would happen if they had the *Enterobacter*-infected mice eat a diet that more closely mirrored the high-fat diet the patient had been eating. Essentially, they drove their furry little companions over to McDonald's and gave them a double cheeseburger, large soft drink, and fries—lots of fat and lots of sugar. And, surprising to absolutely no one, the mice got fat.

But here's the fascinating thing: Per basic scientific procedure, the scientists also kept a control group of mice that consumed the exact same high-fat diet as their counterparts but weren't infected with *Enterobacter*. And those mice stayed skinny as rails.[12]

So was the obese man's diet the problem? Certainly. But that, by itself, may not be the only reason why he was so heavy.

With time we may come to appreciate how genetics, diet, plus the presence of a specific combination of microbes could be helping us tip the scales.

Now we certainly can't "catch" obesity—but we can spread bacteria. And if that type of bacteria is one that potentially contributes to an unhealthy reaction to fats, then the effect could be the same.

But it's not just weight gain that we need to be thinking about when it comes to the effect our personal microbiomes—the menagerie of microbes and their DNA that inhabit our bodies inside and out—are having on our health. It's also our hearts.

You've likely heard that red meat and eggs are bad for your cardiovascular system. What you might not know, though, is that it's not only the saturated fat and cholesterol alone, which we've long assumed causes an increased risk of heart disease. Rather, the risk may be heightened by a compound called carnitine that is prevalent in those foods. By itself, carnitine doesn't appear to be at all harmful. But when met by the bacteria that make up the microbiome living in most people's guts, it's turned into a new chemical compound called trimethylamine N-oxide, or TMAO, that, when introduced to our bloodstream, appears to be bad for our hearts.[13]

So far, the health effects that can be caused by microorganisms that make up the human microbiome have gotten a lot less attention than the human genome. This is going to change as it becomes more apparent, that one's microbiome is just as important as what one eats and the genes one gets. Even monozygotic twins with identical genomes don't always have identical microbiomes, especially when they don't weigh the same.

Which is why, as we're learning about the importance of being stewards of our genetic inheritance, it might be sensible for us to take more of an interest in the welfare of our microbiome as well. One of the easiest ways for us to do that is to consider alternatives to the indiscriminate use of antibacterial products like soaps, shampoos, and even toothpaste. Also, it would be prudent to discuss with your physician the absolute need for an antibiotic prescription before rushing to fill it. As we've learned time and time again, a political regime change that is accomplished by force, or a microbial one caused by medication, can often have unforeseen and long-standing consequences.

GIVEN THE complexity of it all, it wouldn't be unreasonable to give up on trying to wrap your head around where to go next from here.

But let me suggest why there's good reason to be excited about what we're learning about ourselves and our diets—and where that genetic information will take us. And doing that means going back to the emergency department, where Cindy and her mother were already waiting when I arrived shortly before 4:30 a.m.

The staff had started the intake process, and I was glad to see that Cindy already had an IV line running into her arm, bringing her the extra glucose and fluids she so desperately needed. Giving glucose to Cindy is crucial since her OTC deficiency will cause her ammonia levels to rise when she's using protein as a source of energy. Rising levels of ammonia are harmful to the body, and especially to her sensitive and developing brain. This is what is partly responsible for the accompanying symptoms such as lethargy and vomiting that caused her mother to be so concerned.

One of the reasons that the treatment for OTC is so much more aggressive than it was in the past is that we are now much more aware of the accompanying brain damage that comes from having elevated levels of ammonia. One of the treatment options, especially in severe cases, is "gene therapy with a knife," where patients with OTC are given a liver transplant, specifically a liver that gives the patient a working copy of the damaged gene they inherited.

Thankfully, Cindy's case wasn't severe enough to necessitate a liver transplant. But with the rapid change of treatment options that's occurring, OTC deficiency is not the grim diagnosis it was formerly.

As I waited for the results of her blood work (the blood sample had been rushed off to the lab on ice), I thought about all of the significant changes that have occurred in the way we practice medicine in the past few years. In Cindy's case, we previously would not have known that she had a genetic condition until it was likely too late. Which today highlights the imperative of doctors knowing which tests to order to assess a patient's condition.

When Cindy's lab results finally came back they showed that her body's load of ammonia was not as high as we first had anticipated and that her organs weren't showing any major signs of dysfunction.

That was good news. After finishing up my consultation note and e-mailing the day team to hand over our night's business, I left feeling a bit spent. Maybe three and a half hours of sleep wasn't enough after all.

On my bleary-eyed drive home to get showered and changed, I reflected on the sheer magnitude of the biochemical and genetic mysteries that often overshadow our attempts to understand conditions like Cindy's. Witnessing what these brave children and their families go through day in and day out sparks new ways of thinking that occasionally lead me to new opportunities for clinical research. Odds are that I would surely miss new avenues of exploration if I didn't have the honor of spending some time traveling along with these incredible families on their medical journeys.

And as we're going to see next, it was the development of new methods of screening to find children like Cindy early enough to make a difference in their lives that led us to discover that she was in need of a particular dietary regimen and specialized medical care. To see where we are headed in the field of personalized genetic nutrition, it might help to know where we got started. If you or someone you love was born after the late 1960s, you're likely already a beneficiary of it.

IT ALL began in the late 1920s with another worried mother.

She was a Norwegian woman by the name of Borgny Egeland, and she wanted desperately to help her two little kids. Both of her children, a girl named Liv and a boy named Dag, suffered from severe intellectual disability, though Egeland was convinced that they had been unaffected by the condition when they were infants. Her quest for help led her from doctor to doctor and even to faith healers in hopes of finding someone—anyone—who could help her children, all to no avail.[14]

But fortunately, a physician and chemist named Asbjørn Følling decided to take Egeland seriously. While so many others had written Egeland off, Følling listened intently when he learned of her children's plight—and appears to have been particularly interested when he heard that the children's urine had a strange and very musty odor.

When, at Følling's request, a sample of Liv's urine arrived at the laboratory, it seemed at first to be wholly unremarkable; all of the routine tests were normal. But there was one final test—a few drops of ferric chloride to check for the presence of ketones, organic compounds produced by the body when it is burning fat rather than glucose for fuel. If ketones were present, the ferric chloride test should have changed the color of Liv's urine from yellow to purple. Instead it turned green.

Intrigued, Følling asked for another sample, but this time from Liv's brother, Dag. Again, the ferric chloride test turned the urine green. For two months, Egeland brought the scientist sample after sample of her children's urine—and for two months the doctor worked to isolate the cause of the abnormal reaction, finally settling on a chemical compound known as phenylpyruvic acid.

To see if he was right, Følling worked with Norwegian institutions that served developmentally disabled children to collect additional samples and located eight more samples of urine (including two from sibling pairs) from children that responded in the same way to ferric chloride.

But although Følling had identified the chemical culprit for what would turn out to be thousands of cases of intellectual disability, it would be several more decades before other doctors worked out that the condition was due to an inborn genetic error of metabolism (not unlike Cindy's OTC) that prevented these young individuals from breaking down phenylalanine, a chemical common in hundreds of protein-rich foods.

Indeed, as Egeland had first suspected, her children had been born without any signs of intellectual disability. An inherited metabolic condition, ultimately to be named *phenylketonuria*, or PKU, had caused them to build up phenylalanine in their bloodstream at levels that ultimately became irreversibly toxic for their brains.

Once they'd put that together, scientists developed a special diet that could be administered to those identified with PKU, literally preventing intellectual disability. The only catch was that the children had to be identified and switched to that new diet before they became irretrievably symptomatic.

How to know who has PKU—and early enough to leave nothing to chance? That is a problem that ultimately was solved by a man named Robert Guthrie, a physician and scientist who started his career as a cancer researcher. Guthrie ultimately traveled a professional road very different from the one he first intended, leaving research in oncology to study the causes and prevention of intellectual disability, for very personal reasons.

His son was affected with intellectual disability and so was his niece. But her cognitive impairment could have been prevented.

Because his niece was born with PKU.

Using his cancer research experience to tackle the problem of PKU detection, Guthrie designed a system by which small samples of blood, collected and stored on small cards from the heels of newborns, could be tested for PKU. These cards, which came to be known as Guthrie cards, were put into routine use in the 1960s across the United States and in dozens of other nations in the years to come. Over the decades, they've been expanded for use in detecting many other diseases as well.

It took more than 40 years from the time that Borgny Egeland resolved, against all odds, to find the reason for her children's intellectual disability to the time that Guthrie's tests were in widespread use—and, of course, that development came much too late to help the Egeland children.

How can anyone describe the depth of that tragedy? Nor can we adequately capture the glory of that long, long quest toward a brighter future initiated by Egeland and concluded by Guthrie. For that, I leave you in the capable hands of Nobel and Pulitzer Prize–winning author Pearl Buck, herself the mother of an adopted daughter who appears to have suffered from PKU:

"What has been, need not forever continue to be so. It is too late for some of our children, but if their plight can make people realize how unnecessary much of the tragedy is, their lives, thwarted as they are, will not have been meaningless."[15]

And the Egeland children's tragedy was far from meaningless.

Today, Guthrie cards, and the newborn screening that was developed as a result, have been extended to dozens of other metabolic conditions, another example of how one seemingly rare condition can have broad implications for us all. But even newborn screening isn't a catchall. For some people, only sophisticated genetic testing can uncover the big differences small nutritional decisions can have on our health.

IT WAS a rainy morning in Manhattan in the spring of 2010 when I first met Richard.

He was pretty much bouncing off the examination room walls when I walked in. And that, I'd come to learn, was par for the course for this kid.

Of course, rambunctiousness is very common in 10-year-old boys.

But this boy in particular would have run circles around Max from *Where the Wild Things Are*—and as a result Richard had been getting into a considerable amount of trouble at school.

But that wasn't the reason for Richard's first visit to the hospital. Rather, he was there because his legs hurt.

In every other way, and by all visible impressions, Richard appeared to be a picture of good health. His newborn screening? Perfectly normal. His recent yearly checkup? Spot-on average. He seemed to be in such great shape, as a matter of fact, that it took a while for anyone to recognize that there was something wrong with him—and we might not have known at all if it weren't for the fact that some very good doctors took heed of his repeated complaints, rejecting the easy but very unscientific diagnosis of "growing pains."

Without any other good explanation for the boy's leg pain, the doctors ordered a test of his genes—and that test revealed that Richard suffered from OTC deficiency, the same condition we discussed earlier when I introduced you to Cindy.

You might remember that Cindy's OTC symptoms had resulted in many trips to the hospital. Richard's OTC, on the other hand, expressed itself quite differently—it hardly seemed to impact him at all, other than those rather inexplicable leg pains, which might have been connected to the higher-than-normal ammonia levels in his body.

But Richard's other symptoms, to the extent that they existed at all, were so mild that he and his father had a bit of trouble believing there was anything wrong with him whatsoever. On the day I met him, in fact, there was a foil-wrapped pepperoni stick jutting out of his back pocket, even though Richard and his parents had been repeatedly told that people with OTC deficiency are advised to try to maintain a low-protein diet, since they don't handle high loads of protein very well.

That pepperoni stick was a clue as to why his symptoms wouldn't resolve.

What Richard's family didn't realize was that the reports of his lack of concentration at school and at home weren't exactly behavioral but physiological. Higher-than-normal levels of ammonia in most people's bodies can lead to tremors, seizures, and coma, but in Richard, it was likely that his elevated levels were prompting combativeness and difficulty concentrating.

But I'll be very honest—I didn't see this at first, either. Richard had gone home from our first meeting with instructions to stick more closely to his diet because we figured that might help his aching legs.

The only way anyone really knew that Richard's problems were more than skin-deep was when he returned, three months later, this time having adhered much more strictly to his diet. His legs no longer hurt—and that was good—but the big surprise was that he was doing exceptionally well in school. He was calmer. More attentive. He was no longer king of the Wild Things.

I thought a lot in the following months about the implications of Richard's remarkable turnaround. There are, no doubt, more Richards out there. In fact, it's likely that there are many, many more—and they're also eating, quite unwittingly, foods that aren't quite right for their genetic selves. Maybe their conditions aren't severe enough to send them over a metabolic cliff, but perhaps just enough to warrant a trip to the principal's office.

The fact that the children I see are, for the most part, in very specialized medical centers makes me wonder how many patients with metabolic conditions we're missing in primary care—and how many aren't coming in for care at all.

We really don't know how many people who have been diagnosed with some form of cognitive impairment, or even autism spectrum disorder, actually have an underlying metabolic disease that has simply never been diagnosed and addressed. Before we understood PKU, for example, we couldn't understand that these children's intellectual disability was due to an untreated metabolic condition.

The more our science advances, I hope, the more cases like Richard's we'll come to understand—and the more lives we can improve with medical interventions and simple life changes that address people's individual genetic and metabolic needs.

SO WHAT can Cindy, Richard, and Jeff teach the rest of us about nutrition? The answer is that we're all individuals when it comes to our genomes, and completely unique when it comes to our epigenomes and even our microbiomes. Optimizing what we eat is not the same as preventing nutritional deficiencies. We can and should investigate our genes, and metabolism, for clues about what foods suit us best. The findings would have significant implications for what we should and should not be eating.

We are at the point of moving beyond creating specialized diets for people with rare genetic conditions. Because of the information we are now privy to through genetic sequencing, we are now on the cusp of finally being invited to sit down to a meal that's been prepared with our own individually inherited genetic profile in mind.

As we're going to consider next—it's not just our diets that are becoming much more personalized to our genetic inheritance—it's time we looked in our medicine cabinets as well.

CHAPTER 6

Genetic Dosing

How Deadly Painkillers, the Prevention Paradox, and Ötzi the Iceman Are Changing the Face of Medicine

Each year many thousands of people die—and many more become acutely ill—precisely because they were taking the exact dosage of medications prescribed to them by their doctors.

It's not that their doctors were negligent. In fact, in most cases their prescriptions were exactly in line with recommendations provided by drug manufacturers and professional medical societies. The reason for many of these adverse drug reactions lies in our genes. Just like metabolizing caffeine, some of us are just genetically endowed to be better at breaking down some drugs than others. It's not always the version of the genes themselves that you've inherited that can result in adverse drug reactions. Rather, it's also the number of copies of a gene you've inherited that can be just as important. Some of us have inherited a little more or a little less DNA than others, and as you can imagine that sets us up for a lot of variation between people. It's impossible to know what you've inherited unless you get genetic testing or sequencing done to find out.

If you happen to have a deletion in your genome that results in missing sections of DNA that house information that is crucial for your development or well-being, then more likely than not the genetic change can cause a specific syndrome. But when there's a duplication of DNA material it's not always clear what the implications can be.

Having a little extra DNA sometimes has no effect at all, while at other times, it can profoundly change your life. As we're about to see, a little extra DNA can even make a common medication turn deadly. What you've likely clued into by now is that what you do with your genome is just as important as the genes you inherit. And these lifestyle choices include what medications you take.

In one heartbreaking case, a young girl named Meghan died after a routine tonsillectomy, and not because her body couldn't handle the anesthesia or the surgery. In fact, the surgery was a success and Meghan was sent home later the next day. The reason Meghan died was that her doctors didn't know something about her that was vitally important. No one looked at Meghan's genes.

Now, there's a good chance she might have lived her entire life without ever knowing of any differences in her genetic code. What Meghan inherited was a very small duplication in her genome, not unlike millions of other people who have slight differences in their DNA. Because of where the small duplication was located in her genome, instead of getting two copies of the *CYP2D6* gene, one from each parent as we've come to expect, Meghan got three.[1]

And like millions of patients before her, she was given the drug codeine to treat her pain after surgery. But because of Meghan's genetic inheritance, her body was turning small doses of that medicine into big doses of morphine. And fast. The recommended dose that would have ameliorated pain in most children, making them more comfortable, resulted instead in overdose and death for Meghan.

Which is why the U.S. Food and Drug Administration in 2013 finally decided to ban the use of codeine in children after tonsillectomies and adenoidectomies.[2] The tragedy is compounded by the fact that this isn't a rare reaction. As many as 10 percent of individuals of European descent and up to 30 percent of those of North African descent are ultrarapid metabolizers of certain drugs[3] due to the versions of the genes they've inherited.

Given the number of medications we prescribe and the spectrum of genetics involved, codeine use in the pediatric population, is likely only one of many instances in which drugs meant to help people heal are having the opposite effect.

We now have the tools to identify ultrarapid and ultraslow metabolizers of certain medications, including opiates, through relatively simple genetic tests. But there's a good chance that if you were recently prescribed an opiate like codeine in the form of Tylenol 3, you didn't get checked this way.

So why aren't those tests being used more proactively? That's a great question—and one I absolutely urge you to bring up with your physician before you let yourself or your children be treated with certain medications.*

Of course, a risk to some isn't a risk to all. For some people, codeine can be a perfectly safe and effective choice for pain relief.

So what we're moving toward, I hope much sooner than later, is a world where there is no average recommended dose of any drug that is sensitive to your genetic inheritance but rather a personalized prescription that takes into account a myriad of genetic factors and that results in dosages that are just right—just for you.

Besides recommendations for doses of medications that work best for *most* but not *all* people, we're beginning to understand that our genomes also play a significant role in how we respond to preventative health strategies. To appreciate what this might mean to you and

---

* A few of the prescribed medications that are impacted by your genes include chloroquine, codeine, dapsone, diazepam, esomeprazole, mercaptopurine, metoprolol, omeprazole, paroxetine, phenytoin, propranolol, risperidone, tamoxifen, and warfarin.

the health recommendations you're being given, I'd like to introduce you to Geoffrey Rose and acquaint you with his aptly named Prevention Paradox.

SOME DOCTORS are clinicians. Others are researchers. Not everyone can be both—and not everyone who could be both actually wants to be.

But for some doctors, myself included, the chance to see laboratory research reflected in the lives of patients offers incredible opportunities, enormous insights, and the absolute privilege of being in a front-line position to help people.

That's what kept Geoffrey Rose going, too. As one of the world's foremost experts on chronic cardiovascular diseases and one of the preeminent epidemiologists of his time, Rose certainly wasn't required by the research community to do any clinical work at St. Mary's Hospital in London's historic Paddington district. But Rose continued to see patients for decades, even after a brutal car accident nearly claimed his life and resulted in loss of vision in one eye. He kept going, he told his colleagues, because he wanted to ensure his epidemiological theories were always grounded in clinical relevance.[4]

Rose is perhaps best known for his work highlighting the need for population-wide prevention strategies, such as the educational and interventional measures we've applied to the epidemic of heart disease. But he also fully recognized the public health failings of such programs. He called this the Prevention Paradox, which states that a lifestyle measure that reduces risk for the entire population may offer little or no benefit to any given individual.[5] This approach privileges the success of the whole and neglects to tend to the needs of the few who don't quite fit into the rubric of the genetic majority.

Put plainly, the wonder drug for the 5-foot-10, 185-pound white male may not do a thing for you. As we saw with Meghan's prescription for codeine in the beginning of the chapter, it may even kill you.

Even so, we've made incredible gains in health outcomes by treating entire populations with vaccinations such as those that were given against smallpox. However, physicians don't usually treat entire populations but rather individuals within those populations. Yet the guidelines for how we practice medicine are derived from the evidence that is garnered from population studies that are comprised of individuals from an eclectic mix of genetic backgrounds. Which is why codeine was used for so long for pain relief after pediatric tonsillectomies—because it worked on most of the kids, most of the time.

One example of the Prevention Paradox occurs in the first weeks when people with high LDL or "bad" cholesterol start taking fish oil supplements. Researchers found that using fish oil (which is high in omega-3 fatty acids from mackerel, herring, tuna, halibut, salmon, cod liver, and even whale blubber) is associated with a wide range of changes in LDL levels across the population, from down by 50 percent to up by a whopping increase of 87 percent.[6] Researchers have dug deeper to demonstrate that people who supplemented their diets with the so-called healthy fats found in fish oil actually had a greater negative change in their cholesterol levels if they were carriers of a gene variant called *APOE4*. Meaning that supplementing with fish oil may be good for some and very bad for other people's cholesterol levels depending on which genes they've inherited.

Fish oil is by far not the only supplement that millions of people are consuming daily worldwide. It's been estimated that more than half of Americans are thought to be popping supplements, to the tune of $27 billion dollars a year in sales, hoping to prevent illnesses and treat diseases in what seems to be a simple and natural way.[7]

And there are not many medical guidelines or recommendations when it comes to supplements or vitamins, which likely is one of the reasons I often get asked if there's any benefit in taking them at all, and, if so, at what dose. My answer usually has the qualifying word "depends" attached. There are many reasons to take or avoid supplements and vitamins. Have you been told that you are deficient in something particular? Do you have a genetic inheritance that requires you to have an increased intake of certain vitamins? Or most importantly, are you pregnant?

When it comes to fetal development there's no better place to appreciate how the significant mix of vitamins and genes can conspire to prevent serious birth defects. To deepen our appreciation, we'll need to take a trip back to the early part of the twentieth century, where there's a certain sneaky monkey I'd like you to meet.

ONE OF the biggest advances in the eradication of birth defects worldwide started with Lucy Wills and her monkey. And it's a great example of how the old model of "what's best for most people most of the time" has been incredibly effectual at saving and improving lives but has also been ineffective at best (and dangerous at worst) for certain segments of the population.

Like many of the bright young doctors-to-be of the generation born just before the turn of the twentieth century, Wills was fascinated by the cutting-edge field of Freudian thought and had been considering spending her career pursuing the science and art of psychiatry. But while training at London University's School of Medicine for Women, which maintained a close relationship with several hospitals in India, Wills received a grant to travel to what was then Bombay to investigate a little-understood condition called macrocytic anemia of pregnancy, which can cause weakness, fatigue, and numbness of the fingers in some pregnant women.[8] Wills quickly learned something about herself: She loved a good mystery.

At the time, all that was known about the cause of macrocytic anemia of pregnancy was that sufferers had bloated and pale red blood cells. But why? Given that the disease seemed to disproportionately impact poor women, Wills suspected it might be related to their diets. In Wills' day, as in our own, people who were poor and underprivileged typically had less access to fresh fruits and vegetables, and that was certainly the case with the Indian textile workers Wills went to study.

To test her hypothesis, Wills tried feeding pregnant rats a diet similar to what the textile workers were eating. Sure enough, the rats began showing similar changes to their red blood cells, and Wills soon found she could get similar results in other lab animals, too.

With that, Wills began to "build back up" the animals' diets, in much the same way as modern parents are encouraged to introduce new foods to their infants, one by one, in order to make adverse reactions easier to pinpoint.

Wills knew that a complete healthy diet would likely eliminate the problem, but she also knew that she didn't have the power to make that happen for every woman in India. What she needed to do, then, was to identify the exact dietary element that was missing from the women's diets so that it could be supplemented during pregnancy. Despite considerable efforts, though, that exact element remained elusive—until one fateful day, when one of her test monkeys got its hands on some Marmite.

If you're British or live in a country that's a part of the former British Empire, you probably know about Marmite—a sticky, salty, dark brown paste with a love-it-or-hate-it taste made from concentrated brewer's yeast—and its many brand incarnations, including Vegemite, Vegex, and Cenovis. It's certainly not for everyone, but some folks won't leave home without it. Marmite was a staple in British military rations through two world wars. When it ran low in army food supply chains during the conflict in Kosovo, back in 1999, soldiers and their families staged a successful letter-writing campaign to get it back on mess tent tables.[9]

Wills took meticulous notes on everything she did, but there's no record whatsoever of how, exactly, the monkey got its hands on some Marmite. Monkey business being what it is, it's possible the mischievous little creature just stole part of Wills' breakfast.

"Tar in a jar," as it's both fondly and derisively known, is also chock-full of folic acid. And that, Wills discovered as her monkey staged a remarkable medical recovery in the wake of its Marmite feast, was the secret to curing macrocytic anemia of pregnancy.

It took another two decades for researchers to understand exactly why folic acid was such a powerful cure. Since then we've learned that it is crucial for cells that are rapidly dividing, which explains why women who don't get enough of it during pregnancy might turn anemic: Their babies are consuming all their folic acid to grow.

In the 1960s, a connection was also made between folic acid deficiency and neural tube defects, or NTDs—abnormal openings in the central nervous system, such as appear in sufferers of spina bifida—that can run the gamut from relatively benign to deadly. This is the reason that physicians often recommend folic acid supplementation for women of childbearing age even before pregnancy, because the crucial window for its ability to protect against NTDs is in the first 28 days of gestation, a time when many women do not even know that they're pregnant. Folic acid is also associated with reductions in preterm births, congenital heart disease, and, according to one recent study, possibly even autism.[10]

Now, even knowing this, if you still can't bring yourself to spread a glob of Marmite on your breakfast toast, don't worry—folic acid is also naturally found in lentils, asparagus, citrus fruits, and many leafy greens.

The American College of Obstetricians and Gynecologists recommends that all fertile women get at least 400 micrograms of folic acid a day. But that amount is based upon the *average* woman, with *average* genes. And as we know, there's really no such thing as the average patient.

The recommendation also doesn't account for one of the most common genetic variations out there. About a third of the population have different versions of a gene called methyltetrahydrafolate reductase, or *MTHFR*, which is extremely important in folic acid metabolism in the body.

What we don't understand is why certain women who have been diligent in taking supplemental folic acid before they conceive are still having babies with NTDs.[11] It seems that for some women with certain mutations in *MTHFR*, or other related genes involved in folic acid metabolism, 400 micrograms of folic acid simply isn't enough. Because of this, they'd likely benefit from taking even more folic acid, which some physicians are now recommending they do, especially in trying to prevent recurrence of a NTD.

Thinking that it might just be better to be safe than sorry?

Before you run out to the drugstore, though, you might want to take something else into consideration. Taking too much folic acid can mask a different problem, a deficiency of cobalamin, or vitamin B12. In short, seeking to head off one problem could hide another. And as we're still in the very early clinical stages of understanding the short- and long-term risks associated with taking large doses of supplemental folic acid, a better safe than sorry approach might, in fact, be applied by not introducing additional chemical compounds into your body unless you know for certain that you and your baby-to-be need it. Which is exactly why getting a thorough look at your genome would definitely help.

Up until recently, though, there hadn't been a good way to know which version of *MTHFR* people carry. Now there is. Testing for the common versions or polymorphisms in the *MTHFR* gene is now available and is being included in some types of prenatal testing. These screens, or carrier testing, look for thousands of mutations in a few hundred genes. If you're thinking about getting pregnant, it's a good thing to add to the long list of questions to ask your doctor.

Don't be surprised, though, if your doctor doesn't have an immediate or authoritative answer on the availability of commercial prenatal genetic testing for different versions of genes such as *MTHFR*. As the cost of testing has plummeted, there's been a considerable lag between availability of testing and what to do with the information obtained.

In particular, many physicians are still trying to determine the proper steps for effectively counseling women about individualized care—they simply haven't had to do it before. But as doctors learn more about all the different genes we can inherit, such as *APOE4*, and all the things we can do to impact those genes during our lifetimes, such as taking fish oil, things are changing. And fast.

The importance of many of these findings has led to the creation of new fields such as pharmacogenetics, nutrigenomics, and epigenomics, which are aiming to bring together a better understanding of how our lives are both affected and changed by our genes.

Now that you know that genetics plays a role in your nutritional needs, there's one more thing you may want to consider before reaching for your next supplement.

Permit me to take you on an important side trip to explore where our vitamin supplements come from.

MAYBE YOU'VE been on a health kick, or maybe it's a New Year's resolution, or maybe you've just reached that point in your life when you feel it's time to make a change. Or maybe all this talk about nutrition is making you think about your weight, and so you're trying to shed a few pounds or attempting to get a little more sleep. Whatever your plan, there's a good chance that you've either considered using or are already taking a vitamin or herbal supplement.

Or two. Or three. Or seven.

But have you ever wondered about the origins of all those tablets and capsules? Where does the vitamin C in that adorable little chewy bear come from?

I'll bet some of you just said "an orange."

And that's not surprising. After all, the companies that market these products often use oranges and other citrus fruits on the labels of their C vitamins, as if their employees woke up this morning in a Florida orange grove, picked a few plump, juicy fruits off a tree, and, through some magical process, shrank each one of them to look like an edible teddy bear.

The truth is, though, many of the vitamins you and your children might have taken this morning have been created through a process very similar to prescription drug manufacturing. And by one way of thinking, that's good. Consistent manufacturing processes for vitamins and supplements mean you generally are getting the same thing today that you got yesterday, and that you'll definitely get the same thing again tomorrow.

Indeed, besides differing streams of government regulations, the only real difference between prescription drugs and many vitamins is that the latter are based on chemicals that are usually naturally found in food.

But that's not the same thing as ingesting vitamins that are *in* food. Because when we eat an orange, we're not just eating a fruit made purely of vitamin C but rather something that is composed of fiber, water, sugar, calcium, choline, thiamine, and thousands of phytochemicals that are not limited to that single vitamin.

In this way, taking vitamins is a little bit like listening to just the piano loop from "Empire State of Mind." Without Jay-Z's staccato rhymes, Alicia Key's supporting vocals, the rhythm tracks and guitar riffs, you'd be left with no more than the same few measures of repetitive keyboard pounding.

What's missing is the entirety of symphonic nutrition—all of the other phytochemicals and phytonutrients that are in a *real* orange, the purposes of which, as of yet, we don't even fully understand.

That's not to say that vitamin supplementation can't be helpful in certain circumstances, as we've already seen with the use of folic acid for the prevention of neural tube defects. But if you are taking supplements, or giving them to your children, instead of ingesting something you could obtain so much more naturally then you may be missing the true nutritional majesty of consuming vitamins in their most natural form.

Now, if you're committed to applying the latest in nutrigenomics and pharmacogenetics research into your daily regimen, where to begin?

Well, to start, as we've discussed before you should look to learn as much as you can about your own genetic inheritance. You may even consider getting your entire exome or genome sequenced. It's much better to access and use your genetic information while you're still living, though being alive is not really necessary to get results. As you're going to see, when it comes to your genes, even the dead can speak.

THE BODY was disfigured and terribly decomposed. So when a small group of hikers stumbled upon it while trekking through the Ötztal Alps, near the border of Austria and Italy, they initially assumed that they'd discovered the remains of another mountaineer—perhaps someone who had died several seasons back.

It took several days to get the body down from the mountain, but once that happened, it was clear that this was no ordinary hiker. Rather, the corpse was an exceptionally well-preserved mummified body that was thought to be at least 5,300 years old.

In the decades since Ötzi's discovery, we've learned a tremendous amount about his life and death. For starters, it appears that he was murdered—his violent demise seems to have been caused by an arrowhead lodged in the soft tissue of his left shoulder and a subsequent blow to the head. Analysis of his stomach and intestinal contents show he'd eaten well in his final days—Ötzi had dined on grains, fruits, roots, and several types of red meat.

But it wasn't until researchers removed a tiny piece of bone from Ötzi's left hip that the real genomic fun began. Genetic analysis of DNA preserved in the bone showed that although Ötzi was discovered in Italy's frigid mountainous north, it seems his closest living genetic relatives today are islanders from Sardinia and Corsica—more than 300 miles away. He also likely was light-skinned, had brown eyes, had type O blood, was lactose intolerant, and was at increased genetic risk for dying from cardiovascular disease, meaning that if we could go back in time and keep him away from milk, meat, and murderers, Ötzi might have lasted a bit longer than his estimated 45 years.[12]

For Ötzi, it's a little too late for any of that genetic information to help. But if we can discover that much about someone who died roaming the Alps more than 5,000 years ago, just imagine what we could learn about ourselves today.

For those who may not have access to comprehensive genetic testing and sequencing, there's still another low-tech option that doesn't require you to subject yourself to the same sort of posthumous rigorous genetic testing that Ötzi endured. A routine climb up your family tree can help you get a lot of valuable information. Asking your relatives if they've ever had an acute drug reaction, for example, just might save your life.

And when trying to parcel out a complex disease that results from myriad genetic interactions, *any* bit of information can be crucial. The truth is, there's really no replacement for a good family medical history. And that's why, when it comes to genetic health in the decades to come, the Mormons may be leading the way.

You might know Mormons as members of the fast-growing international Church of Jesus Christ of Latter-day Saints. And you might have occasionally encountered them directly—in teams of two, their hair cropped and gelled back, dressed in dark slacks and white shirts with black name tags—at your doorstep.

What you might not know, though, is that some Mormons also engage in a practice known as baptism for the dead, under the belief that people who died without having had the opportunity to be baptized by a proper authority can get a second shot at salvation, so to speak, if they receive a proxy baptism from a living Mormon.

That rite gave rise to the modern Mormon practice of sophisticated, computer-based genealogy research, which is a key reason why many members of the church can recite the names and life stories of their ancestors going back for hundreds of years—even for lines of the family complicated by a single husband and multiple wives. This is to ensure that no Mormon soul is left behind.

For doctors trying to link genetic conditions with family histories, that kind of detailed information can be an absolute gold mine. Today, the church makes many of its genealogy records available to the public over the Internet,[13] and many non-Mormons take advantage of that, but for church members, it's literally something to be done religiously.

And since Mormons have long sustained a rather strict set of guidelines about what they put into their bodies (many don't drink caffeine, most eschew alcohol, and illicit drugs are particularly shunned), they might have fewer complicating factors to deal with when sorting through the genetic, epigenetic, and environmental issues at play in their lives.

YOU DON'T have to be a Mormon to give your siblings, children, and grandchildren a better shot at having the important information they'll need to make better sense of their genomes, and thus their personal health. One of the best gifts you can provide them with is a thorough genealogical history, starting with what you know about the health of your own parents and moving on up and across the family tree as far as you can.

Make it as detailed as possible: You never know how some seemingly inconsequential detail of one generation, like sensitivity to a specific drug, can lead to a useful bounty of familial medical information. So knowing more about your own inheritance, either through a detailed family history or from direct genetic testing, can serve as an important reminder about your own unique individuality.

It's a reminder that tells you that it's time to step away and leave the crowd behind by starting to ask questions like these: What's the best drug and dosage for my genotype? How can I avoid the Prevention Paradox? What nutritional and lifestyle strategies should I be trying to employ to best serve my genetic needs? And what genetic life lessons can I learn from a frozen 5,000-year-old Italian mummy?

You may not find all the answers to these key questions right away, but by asking them you'll come closer to getting a picture of some of the most important genetic qualities that make you incomparably original.

CHAPTER 7

Picking Sides

How Genes Help Us Decide Between Left and Right

The raging bull was done. He'd been put out to pasture. That's what they said.

And it wasn't just the critics—although there were many of those too. It was fellow surfers. They'd known for a long time that Mark Occhilupo's demons were getting the best of him. They knew the drugs had taken their toll. They could see him getting bigger around the waist and falling further and further behind the other top surfers of the day.

In 1992, it all came to an explosive apogee. At the Rip Curl Pro competition on southeast France's famed Hossegor Beach, the man known around the world as Occy reportedly attempted to push over the judges' booth, threw a board at his opponent, and even chowed down on some beach sand before announcing he would be swimming home to Australia.[1]

The self-assured, swaggering Aussie had never won a world title. And when Occhilupo abandoned the Association of Surfing Professionals championship tour that year, it seemed clear he never would.

Out of the limelight, though, Occy set to work righting his life. He got sober. Got back into shape. He swore off the fried chicken, which had been a staple of his diet for way too long. He started surfing again, this time for fun and fitness rather than fame and financial gain.

Then in 1999 Occhilupo grinded his way, wave by wave, win by win, to the Association of Surfing Professionals World Tour Title. At 33, he was the oldest champion ever.

Years later, Occy was still at it. After yet another retirement—this one came about on easier terms than the first—the raging bull was raring for another shot at the world circuit. It was then, on a stunning Hawaiian morning on the island of Oahu, that I watched Occhilupo dive headfirst into the crashing waves, emerging not long afterward over the frothy crest and dropping into the trough with all the effort that any of the rest of us might put into laughing at a great joke.

I'm not a pro surfer, but one thing really stood out to me as I watched Occhilupo ply his trade that day: He's goofy.

Some people call left-handers southpaws. Others call them mollydookers or corky dobbers. Scientists still often call lefties sinister, which in Latin originally just meant "left" but later came to be associated with evil.[2]

Wondering about the medical implications of being born a corky dobber? It may surprise you that left-handed women were found to be twice or more likely to develop premenopausal breast cancer than right-handers. And a few researchers believe this effect may be linked to exposure to certain chemicals in utero, affecting your genes and then setting the stage for both left-handedness and cancer susceptibility,[3] thus opening up another probability of nurture changing nature.

When it comes to our hands, feet, and even our eyes, most human beings are right-side dominant. Now, you might think that footedness and handedness are always aligned, but as it turns out that's not always the case for right-handed people, and it's even more infrequent for left-handed people. Lots of people aren't *congruent*.

In board sports, though, the term is *goofy*, and it refers to which foot is planted toward the back of the board, and thus which foot dominates when it comes to turn control. Occy stands with his left foot back.

There are an amazing number of theories as to why some of us are goofy-footed. But the term itself is often said to have originated with an eight-minute long Walt Disney animated short, called *Hawaiian Holiday*, that was first released to theaters in 1937. The color cartoon stars the usual suspects: Mickey and Minnie, Pluto and Donald, and, of course, Goofy. During the gang's vacation in Hawaii, Goofy attempts to surf, and when he finally catches a wave and heads back to shore atop its short-lived crest, he's standing with his right foot forward and his left foot back.[4]

If you're wondering if you might be goofy and would like to find out before hitting the beach, then imagine yourself at the bottom of a staircase that you're about to ascend. Which foot moves first? If you're taking that first imaginary step with your left foot, then it's likely that you're a member of the goofy-footed club. And if you find out that you aren't goofy, then you're in the majority.

Why some of us are born left-handed, right-handed, or goofy-footed is thought to relate to an important and early time in the formation of our brain. One of the most popular explanations for *lateralization*, which is the term given to this phenomenon, is that each side of our brain has evolved for functional specialization. This division of labor allows us to perform multiple complex tasks.

Do you whistle while you're at work? Your coworkers can thank your brain's remarkable lateralization for that. Are you able to drive and talk on the phone at the same time? That's lateralization, too.*

---

* You might not be as good at it as you think—research has shown that cell phone users are generally as bad as intoxicated drivers when they get behind the wheel of a car.

So why the predominance of righties? For our species, one of the most important tasks is communication, which is generally processed on the left side of the brain. And some scientists think that's the reason why we're right-side dominant, because, as you've probably heard, the left side of the brain generally controls the muscles on the right side of the body (which is why a stroke suffered on the left side of your brain is more likely to result in impairment to the arm and leg on the right side of the body).

So why should you care if you're goofy? Well, it's the same question that many people have posed to Amar Klar, a senior investigator of the Gene Regulation and Chromosome Biology Laboratory at the National Cancer Institute. He has been interested in the genetics of handedness for more than a decade.

Klar is a proponent of a direct genetic cause for handedness, perhaps even a single gene—a discovery we've thus far managed to miss as we've combed the human genome. The theory, which Klar's team has backed with a predictive model of dominant and recessive traits that would make Gregor Mendel proud, even explains the fact that monozygotic twins don't always share the same handedness. This might seem to be an argument against genetic inheritance, but what Klar and several other respected geneticists have proposed is that this theoretical gene carries two alleles, a dominant one that orders up right-handedness and a recessive one. Someone who inherits a pair of recessive alleles has a 50-50 chance of going either way. More than a decade after he started looking for that elusive gene, Klar hasn't found it yet, but he's still holding out hope.

As an alternative to an exclusively genetic cause of handedness, a different line of thinking suggests that left-handed individuals experienced some neurological insult, or damage, during development or delivery that affected the way their brains are wired.

Marshaling evidence for "the insult theory," some people have pointed to studies that found a correlation between children who are born premature and left-handedness. A Swedish meta-analysis[*] found an almost twofold increase in left-handed children who were born premature.[5]

Discovering more of the biology behind handedness, tracing it to genetics, exposures, or both, could give us a lot more knowledge as to whether we should line up our kids on the left or right side of the tee-ball batter's box. That's because left-handedness has also been associated with higher rates of dyslexia, schizophrenia, attention deficit hyperactivity disorder, some mood disorders, and, as we've discussed, even cancer.[6] Indeed, adding handedness into the mix has helped Danish researchers identify which children who had symptoms of ADHD at the age of eight (when, let's face it, just about every kid is a little bit on the rambunctious side) would still have it at the age of 16.[7]

Unlike handedness, though, we're a lot closer to understanding the genetic reasoning behind the anatomical planning that happens during the development of our body—the genes that work hard to ensure our hearts and spleens end up on the left and our livers on the right. This genetic understanding helps us answer the following question.

DOES IT really matter which side does what? If you've ever experienced the joy of a hot tap marked cold, then you've experienced the pain of laterality gone wrong. When our bodies don't work as labeled or expected, things can get dangerous—or at least a little goofy.

---

[*] A meta-analysis is a study that combines the results of many similarly designed studies to increase the statistical power and thus the accuracy of the results.

But first, to really understand how genes help your body pick sides, we're going to need to travel back in time to when you were just starting out your life's adventure as an embryo in your mother's womb. As we begin our development in three dimensions, there's an exquisite balance of growth that needs to be struck to make sure we can maintain what will become our future bodily proportions.

The funny thing about imbalance is that it doesn't take a whole lot of it to throw everything out of whack. So while a little biological one-sidedness might be good for life, just a little more can cause things to go seriously awry. And quickly, too.

If you've ever been on a small boat—a canoe, perhaps, during a camping trip—you know how this goes. With everyone seated and rowing in perfect coordination, a canoe is an incredibly stable way to move across the water. But all it takes is for one person to stand up at the wrong time and the whole thing capsizes.

I thought about this as I stood on the beach on Oahu's North Shore and watched Occhilupo burst out of the barrel of a wave as it crested to the left, then cut back sharply, always one step ahead of the break, manipulating the water like a Japanese chef cutting up a piece of chicken breast sizzling on a teppanyaki.

Occhilupo's a master craftsman, but even he couldn't have done that if not for something that happened back in the 1930s.

If you watch that *Hawaiian Holiday* cartoon, you might notice that Goofy's surfboard looks a bit like an ironing board. It's long, flat, and tapered at one end—and it has nothing on the bottom. That's because Goofy's board hadn't yet met a guy named Tom Blake, an inventor and surfboard maker who, just a few short years before that cartoon was made, had introduced the world of surfing to the skeg, a fin attached to the underside of the board that helps maintain balance and provide improved maneuverability. As the story goes,

Blake's first prototype was part of a motorboat's keel that had washed up on shore.

At first no one really understood what good such an appendage on a surfboard would do. Within a decade, though, nearly every surfboard in the world was outfitted with one or more fins.[8]

How does surfing relate to genes and our own development? We humans don't have a skeg, per se, but a similar sort of structure coded deep within our genes plays an absolutely vital role in our development and sets up the environment for the right genes to be expressed at the right time. Chances are, though, you've never heard of them. They are called nodal cilia, and they show up during embryonic development—at a point in which we more or less resemble a squashed piece of gum within our mother's uterus. At that all-important juncture, nodal cilia stick up from what will be our heads, like little protein antennae.

And just like a skeg that helps a surfer steer his surfboard in the water and shred some decent waves, our cilia are crucial for moving (and in some situations sensing) the fluid around our developing embryonic selves and creating a necessary spatial chemical concentration gradient. In this way, cilia are simple but vital: Moving the fluid in a specific direction, creating a current like a whirlpool around the embryo. That changes the amount of proteins that are floating in just the right order, which then directs your body's development, through genetic expression, at just the right time.

Our developing embryo uses these protein signals, which are encoded by our genes, to make sure that our liver forms on what will become the right side of our body and that the spleen will develop on the left.

In the grand battle fought between competing sides of a human body as to which side gets which organ, our genes code for aptly named proteins like Lefty2, Sonic Hedgehog, and Nodal that duke it out for supremacy in the realm of laterality.

But when cilia are not working well due to a genetic change, our developmental balance can go completely awry. Like a surfer whose skeg has been broken off by an offshore reef or an unexpected tide swell, misbehaving cilia can cause an imbalance in the amount of proteins that wash over the embryo.

And if more of the protein Sonic Hedgehog flows beyond its usual borders, it can, metaphorically speaking of course, eat your spleen, leaving you spleenless. Not to be outdone by Sonic Hedgehog, when proteins like Lefty2 are not working, you can end up with more than one spleen, a condition we call polysplenia.

Confused cilia can even flip our organs goofy. Spin the whirlpool the wrong way and you can end up with some of your major organs on the complete opposite side of the body—the heart on the right, the liver on the left, the spleen on the right.

Far from being benign, if the proper placement of our internal organs gets lost during development, it can affect almost everything, from our vascular plumbing to our neurological wiring. And what has been done anatomically and neurologically cannot easily be undone. Often, it cannot be undone at all.

That's why obstetricians stress avoiding alcohol during pregnancy. For the most part it's been assumed that when it comes to alcohol and pregnancy, there is no known safe level of exposure. On the other hand, though, we know that sometimes babies are born to mothers who drank alcohol during their pregnancies and those children appear virtually unscathed.

Why the difference? Because we are all genetically diverse—and particularly, it seems, as it relates to alcohol metabolism. Depending on what genes have been inherited by a mother—and what genes she and her partner have passed on to their child—the impact of alcohol on a fetus can be mildly toxic or like an incredibly potent straight-up poison.[9] Given the uncertainties during this part of our children's developmental journey, the best approach, in my opinion, is still to avoid drinking during pregnancy altogether.

That's probably good advice for any questionable substance, including unhealthy foods, that a woman puts into her body during pregnancy, but it might be especially important when it comes to alcohol—and particularly during the first stages of development, when sober cilia, so to speak, are vitally important.

In a way, cilia are sort of like genetic conductors in a developmental orchestra. If you've ever watched an orchestral maestro at work, you know that it's hard enough to make symphonic music when you're sober. Just imagine trying to do it while you're drunk. That's why researchers have found that children of mothers who drank excessively during pregnancy may have many issues related to laterality, including difficulty hearing from their right ear and challenges interpreting speech, both functions that are generally processed on the left side of the brain.[10]

Instead of genetically directing the developmental orchestra through a spectacular performance of harmonies, melodies, and rhythm, malfunctioning cilia conduct performances that are more reminiscent of the work of Japanese composer Toru Takemitsu, whose often discordant compositions are fascinating to contemplate and study but can be hard to understand. And that's the challenge with genetic diseases known as ciliopathies, which are caused when cilia fail to perform their normal functions.

To understand ciliopathies, it's important to understand cilia and the genetics that are behind them. And to do that, first you must know that cilia are everywhere—and I mean absolutely everywhere. While you might never have heard of them, they've been looking out for you and your well-being since before you were born. Like a modified form of touch, some of your cells even use cilia to physically sense their way around their microscopic world.

However, there are other compelling examples of the importance of using touch to make sense of the world around us.

THE AMERICAN sculptor Michael Naranjo was blinded and lost the use of his right hand in a grenade attack when he was a 22-year-old soldier in Vietnam. While being treated at a hospital in Japan, Naranjo, who came from a family of artists in New Mexico, asked a nurse if she might find him a small piece of clay. A few days later she was able to fulfill his request, and Naranjo set out on an artistic journey that has taken him around the world.[11] Many years later, he was even invited to the Galleria dell'Accademia in Florence, Italy, where a special scaffold was erected so that he could run his hands over the face of Michelangelo's *David*. This is how Naranjo sees.

Like this amazing artist, our cells are physically blind and use their genetically encoded cilia as a means by which to sense the world around them. Even though cilia are so fundamental to our own lives, given their hidden microscopic size most of us don't give them a second thought. What they lack in size, they more than compensate for in consequence.

Their impact on our lives begins very early—earlier even than the time that our cilia get to work stirring and sensing the embryonic fluids that make us who we are—because cilia also play a vital role in conception.

For starters, the tail of a sperm is a modified cilium known as a flagellum. If it doesn't beat right, it won't swim right, and if it doesn't swim right, it won't get to where it's supposed to go. On the other side of the operation, cilia sit at the entrance to the fallopian tubes, where they beat faster during ovulation to create a strong current to usher in the egg from the ovary.

Our lungs are also considerably dependent on cilia to keep things physically tidy, and this is an important factor that helps oxygen move from the outside world into our bodies. Like concert revelers passing a crowd surfer across a sea of outstretched arms, our cilia also clear out mucus, dust, and microbes from our lungs. That's a tough task even in the best of circumstances, but it's made all the more difficult when we smoke, inhaling chemicals that can adversely affect cilia. Anytime you hear a smoker's cough, you can say a little thank-you to your cilia—because that's what we'd all sound like if those genetically driven little guys weren't doing their jobs.

But you don't have to be a smoker for this process to break down. All you have to do is inherit specific mutations in genes, such as *DNAI1* and *DNAH5*, which cause cilia to misbehave. The genetic condition caused by mutations in these genes is known as primary ciliary dyskinesia or PCD. As we're beginning to understand, more and more, most of what cilia are doing remains hidden to all of us. But when they're not working well, the muscle and elastic tissue of the lungs eventually break down, resulting in difficulty breathing and swollen sinuses that block nasal drainage. All of these symptoms are the result of genetic conditions involving cilia that, for one reason or another, haven't gotten the signal to beat the way they're supposed to.

Some people with PCD also can have *situs inversus*, which, among other things, creates a great opportunity for senior physicians to have a field day with young doctors.

I went through this hazing ritual once while I was a medical student. During an observed physical examination, one of our physician instructors asked me to "tap out the liver." This a percussive technique that has been used for centuries by doctors to estimate the size of this vital organ—something that is crucial to know, even today after the advent of ultrasound. But the senior doctor conveniently didn't mention before I started that this particular patient had *situs inversus totalis*, which means all of her major organs were on the side opposite of normal.

"Moalem, is there a problem?" the doctor asked as I fumbled around the patient's abdomen, trying desperately to repeat what I had practiced so many times on friends, family, and patients while studying for my examinations.

"Well...I...um..."

"Come now, lad, just tap it out."

"I am...I mean...it seems as though...um..."

At this point I was so flustered that I didn't notice that the patient, who was in on the joke, was trying hard to control her laughter. She finally began laughing hysterically—a sign I at first took to mean that I must have been inadvertently tickling her abdomen as I searched for her seemingly absent liver. It wasn't until everyone in the room began laughing as well that I realized that I was, in fact, the butt of the joke.

Now, looking back, I can easily say that this particular practical joke, while quite embarrassing at the time, was one of the most instructive lessons of my medical education. It taught me to always take a moment, before examining a patient, to clear my mind of any assumptions I might have.

TURNING A physician's mind into a medical tabula rasa isn't easy. Some things we simply take for granted—especially if, as part of our medical training, we come to have certain clinical assumptions about human anatomy and physiology.

Indeed, this has become even more challenging as I have become a busier physician. But it has also become more important, because the closer we get to truly personalized medicine, the more critical it becomes for us to move beyond previous assumptions.

There are still some things, though, that we believe to be true for everyone. When it comes to our health, the genetics behind our cilia are unwaveringly important. Helping embryos decide where to form their internal organs isn't all that cilia do. They are also involved in the proper internal structural formation of our kidneys, liver, and even the retinas of our eyes.[12] Just like Naranjo's hands running across a piece of marble, modified cilia even help facilitate proper bone formation as they help cells spatially orient themselves in three dimensions.

As it turns out, there's almost no place in our bodies where cilia have not played a major role. And yet they still remain one of the most fundamentally under-studied structures we have.

Without genes that give us working cilia, we don't have laterality. And without laterality, our internal organs and brain don't form properly. This is why laterality is at the heart of life as we know it. As we're about to see, laterality or sidedness has unspeakably deep genetic implications, ones that may be literally out of this world.

SOMETIMES, WE just have to pick a side. I witnessed a comical real-world example of this a few years back while getting ready to cross a bridge that served as a border crossing point between Thailand and Laos. Thais drive on the left and Laotians drive on the right. When the border crossing opened on that morning there was a substantial bit of chaos and hilarity as drivers tried to figure out which side of the bridge they were supposed to cross on.

It's like this deep within our bodies, too. Without choosing sides, we'd be quickly lost in a world of molecular and developmental chaos. Because of this, almost everything is set up in such a way that it is oriented to the left or to the right. And despite what the "righties" of the world would have you believe, our internal biochemistry seems to favor so-called "left-handed" molecular configurations.

Take the 20 different amino acids that work together to build millions of different proteinaceous combinations. At a very basic level, our bodies use amino acids as the material building blocks that give our bodies form and function. The specific order in which amino acids are strung together is dictated by information that is provided and translated from our genes. A change in one letter of DNA can change the amino acid that is used in making a protein and it can also completely change the protein's ability to do its job. And, of course, that makes amino acids and the order in which they are put together extremely important.

Amino acids (save one exception—glycine) are chiral, which means there can be right-handed amino acids and left-handed amino acids. In fact, when we synthetically create them in a lab, we often can get an equal mix of righties and lefties.

Now, there's nothing wrong with right-handed amino acids. They certainly can behave just like the left-handed ones. If you pile them up one atop the other like stackable chairs, they're just as stable. But for some reason, biology on this planet seems to favor lefties.

Now, if you're thinking that this is all beginning to sound just a little bit out of this world, you're right on track with a theory being worked out by NASA scientists. And it is quite literally otherworldly.

After procuring a few fragments of a meteorite that fell on Tagish Lake in northwest Canada in the winter of 2000, NASA scientists mixed the samples in hot water. They then separated out the molecules bit by bit, using a technique called liquid-chromatography mass spectrometry, a common laboratory process for separating out individual molecules from a mess of other molecules.

Lo and behold, they found amino acids.

But the NASA folks didn't get all starry-eyed there. They kept going. They started sorting the lefties from the righties. What they found was significantly more left-handed amino acids than right-handed ones.[13] The implication, if the research holds up, is that the excess of left-handed amino acids we have here on Earth may have come from a galaxy far, far away. And that might mean that our little corner of the universe itself leans just a little to the left.

LET ME let you in on one of the biggest secrets the supplement industry would rather you not know—some of the vitamins that you are buying and consuming are doing more harm than good. All thanks to handedness. Vitamin E is one such example of this. You might know it as an important antioxidant. Back in 1922, we called it tocopherol, from the Greek meaning "to bring child," since one of the only things we knew about it then was that deficiency in this vitamin led to infertility in rats.

We find vitamin E in a variety of the foods we eat, including leafy vegetables. And yes, it's known as a protectant for the membranes of cells from the chemical onslaught of oxidation, sort of like a rust prevention treatment that might protect the underside of your car from the ravages of weather and road salt. But that's not all it does. We've also learned that it can dramatically change the expression of certain genes, including those associated with cell division—something that must happen millions of times a day to sustain our lives.[14]

Where does the vitamin E used in supplements come from? Vitamin E, like most commercially available supplements today, is made artificially in chemical factories.

The form of vitamin E often found in supplements is alpha-tocopherol, which itself can come in eight different forms, called stereoisomers—only one of which is actually found in the natural foods we eat. And for many decades we've known that, in high doses, alpha-tocopherol brings down the levels of the naturally occurring gamma-tocopherol found in our diets.[15] In other words, the artificial capsule version can counteract one of the naturally ubiquitous forms of vitamin E.

In light of that, might I suggest that you skip the little capsules and cartoon-shaped tablets and instead eat foods that are rich in vitamin E, like certain nuts, apricots, spinach, and taro. Nature, as it turns out, is usually a pretty good arbiter of the types of vitamin E variants we actually need.

Taking our vitamins by eating sensible meals is beneficial in another way. It makes it a lot harder to go way beyond what is reasonable and prudent when it comes to vitamin intake.

And, at this point, I probably don't even need to mention the fact that your specific genotype can have a significant impact on how you metabolize individual vitamins. In fact, a recent study has even identified three different genetic variations that impact how men respond to vitamin E supplementation.[16]

But the key for most of us is simple equability, in which the equilibrium of our bodies, our lives, and even our universe is dependent on just the right amount of imbalance.

In this way our genes help us choose between left and right. We owe our lives and the normal development of our brains to this well-orchestrated balance of laterality. Without having the right genes being turned on at just the right moment, we'd all be mixed up inside and out, from our spleens to our fingertips.

CHAPTER 8

We're All X-Men

What Sherpas, Sword Swallowers, and Genetically Doped Athletes Teach Us about Ourselves

There's a Coca-Cola machine on the top of Mount Fuji.

That's about all I can recall from my time at the summit of Japan's tallest mountain.

Unfortunately, there's plenty else that I remember from the climb itself, which I began at dusk in the Land of the Rising Sun. It takes most people about six hours to reach the summit, and those who travel at night (as I did, in anticipation of getting to the top with plenty of time to spare to watch the sunrise) are advised to build in plenty of extra time.

But I was young, healthy, and confident that I'd be leaving everyone else in that big, beautiful mountain's volcanic dust. I planned to stop along the way at one of the crowded mountain rest-huts for a hot bowl of udon noodle soup and maybe a quick power nap, then continue on to take the summit in time to create a proud and beautiful memory.

Man, was I delusional.

Getting to my intended resting spot was the easy part, though it did take me quite a bit longer than I'd imagined it would. The higher I got, the slower I went. My legs weren't tired, but my mind was. I knew I'd slept a good eight hours the night before, but I told myself that it must have been a fitful sleep, perhaps owing to my excitement over this much-awaited climb.

Yes, I thought, that must have been it.

Nonetheless, I was determined to reach the summit before daybreak. I skipped my intended *inemuri*—that's what the Japanese call a power nap—slurped down my bowl of udon, filled my metal thermos with hot green tea, and hit the mountain trail.

And then, like a karate master, the mountain hit back. Hard.

I spent most of the rest of the climb fighting rain, then sleet, then pellets of hail. But the weather wasn't the biggest problem—not by far.

My head was pounding. I was nauseated and light-headed. The world was spinning. Imagine the worst hangover you've ever experienced—this was worse than that. I doubled over on the side of the trail, not able to continue and at a complete loss for what to do next.

My mind simply refused to work.

And then, to my rescue, came an elderly Japanese woman. I'd first met her at the base of the mountain a few hours earlier when she'd asked me to help steady her as she tried to get into some oversized foul weather gear. She had proudly pointed to both her hips and left knee, letting me know that she had recently been "upgraded" with stainless steel and titanium implants. Because of this, I had been certain she wouldn't get even halfway up the mountain. In fact, to be honest, given the weather and the difficulty of the climb, I had been more than a little worried about her.

Now, here I was, being helped by a woman close to the age of 90 who was gracefully hobbling up the side of a mountain with the aid of two canes. She stopped to take my pack and helped me to my feet.

I was pretty sure that nothing could be more humiliating. But I was wrong. Much to my own dismay, as well as to the dismay of those around me, I then learned firsthand just how much flatulence human beings are capable of producing.

Yes, I farted my way up Mount Fuji.

I'd heard of hypobaric hypoxia, a lack of available oxygen due to a decrease in atmospheric pressure. But I never experienced it before that night, and my mind was in no condition to realize that flatulence, light-headedness, confusion, and exhaustion were all just part of the joy of altitude sickness.

But why was this specifically happening to me and not to my sweet, elderly climbing partner? Why was she able to keep chatting away, carrying my pack along with hers, and occasionally looking back to flash toothy smiles of reassurance as I fought desperately to keep up?

Well, it turns out that my genes apparently leave me a bit more susceptible than most to altitude sickness. Instead of helping me with my climb up Mount Fuji, my genetic inheritance was weighing me down.

If only I was a little more Sherpa.

ALMOST EVERY civilization has a story about how its people came to be where they are today. Quite often these stories of origin have to do with a physical journey. A trip across a raging sea, a flight across a barren desert, a crossing through a rugged mountain range.

There's good reason for that. Although today we might feel separated by language, culture, or politics, our collective human story is one of movement—a search for greener pastures, a quest for giving seas. And as people travel, so do their genes. Indeed, we are all genetic migrants.

These days, with the help of widespread genetic mapping, we're increasingly able to scientifically explore stories of origin, but there are still many holes to be filled and stories yet to be discovered.[1]

To me, one of the most fascinating stories to emerge is that of the Sherpas, who are thought to have arrived about 500 years ago from other regions of the Tibetan Plateau to a particular place in the Himalayan Mountains. That was the nearest they could get to a sacred peak they called Chomolungma.[2]

You might know it as Mount Everest.

The biggest problem with being so close to the Mother of the World, as the peak is known to the Sherpas, is that the great matriarch exists amid a scarcity of the very substance that makes human life possible on this planet. At more than 13,000 feet in altitude, the Tibetan village of Pangboche, the oldest Sherpa village in the world, is nestled nearly a full mile above the altitude where many people begin to feel the effects of hypobaric hypoxia.

I, for one, am not planning a visit anytime soon.

So, what happens to most people at that altitude? Well, for those who ascend very gradually, perhaps just a little headache, fatigue, nausea, or even euphoria.[3]

But as we're about to see, those who haven't inherited specific genes for high altitude living can suffer the consequences, as I did. Even if you don't have the genetic makeup that makes high-altitude living comfortable, there are a few things you can do. You can take the time to try to acclimatize as you ascend and let your genome, through genetic expression, help you adjust.

Or there are some drugs you can take—some with a prescription and others not. Some South American indigenous groups are said to chew coca leaves to deal with the symptoms associated with altitude sickness. There are also anecdotal cases that suggest that caffeine at high altitudes might be helpful.[4] Maybe that's why the can of Coke I had on top of Mount Fuji tasted so good. Back then I thought it was because I paid ten dollars for the honor of acquiring "a passport to refreshment."[5]

In most circumstances when we spend a lengthy amount of time at higher elevations our genes begin to subtly adjust their expression, which prompts cells in our kidneys to make and secrete more erythropoietin, or EPO for short. This hormone stimulates cells in our bone marrow to increase the production of red blood cells, as well as keep the ones already in circulation around past their typical expiration date.

Our red blood cells normally make up a little less than half the content of our blood, with men having a little more than women. The more we have, the better we are at absorbing and shuttling about the vital oxygen our bodies need to survive. Because red blood cells are like little oxygen sponges. And the higher up you are, the less oxygen there is, so the more red blood cells you need. Our bodies' physiology recognizes these changes and signals our genes to shift their expression to accommodate them.

When you need to make more EPO, your body increases the expression of a similarly named gene. That serves as the genetic template for the production of more EPO. However, nothing in your biological life is free. So EPO needs to work a bit like a Washington, D.C., lobbyist, convincing members of Congress to spend a little more capital on the production of red blood cells when your oxygen availability decreases. And, just as in Washington, increased funding for one pet project often comes at the expense of another. Biological currency, after all, is not much different than greenbacks—and like all forms of capital expenditures, there are always some unforeseen costs.

In the case of increased genetic spending on EPO—which causes you to have more red blood cells—another biological cost is that blood becomes thicker. Like high-viscosity motor oil, your blood moves a little more slowly through your system. And that, of course, makes clotting more likely.

As long as things don't get too thick for too long, though, a little extra genetic production of EPO can be just what your body needs to increase oxygen flow. Just as a lack of oxygen can leave you feeling lethargic, a surplus can provide your body the ability to utilize and burn more energy. That's why synthetic EPO has been such a gift for people who can't make enough of their own due to kidney failure, and who suffer from anemia as a consequence.

But that's also what made synthetic EPO a darling of quite a few people in the professional endurance sports community. That is at least until tests were developed to detect it. Among those who have admitted or otherwise been caught "doping" with synthetic EPO are Lance Armstrong, fellow cycling champion David Millar, and triathlete Nina Kraft.

Not everyone has to dose up on synthetic EPO to get a bit of a competitive advantage, though. Take Eero Antero Mäntyranta, for example. The legendary cross-country skier who won seven Olympic medals for Finland in the 1960s is affected by a genetic condition called *primary familial and congenital polycythemia*, or PFCP, which means he has naturally higher levels of circulating red blood cells pulsing through his arteries and veins. And that means he has a natural genetic advantage when it comes to aerobic competition.

So here's a question: If some people have a natural genetic advantage—extra oxygen-carrying capacity in their blood, for instance—is it truly unfair for others to try to bring themselves up to that level? To be clear, I'm not advocating doping. But as we learn more about how our genetic inheritance impacts our lives, we're likely to be confronted with the reality that some of us are genetically doped to begin with.

It would be ridiculous to singularly reduce Mäntyranta's Olympic successes to the genes he happened to have inherited. Even for a biologically advantaged athlete, the level of training required to compete at the international level is extreme. But as with Shaquille O'Neal's imposing 7-foot-1-inch frame and Olympic champion swimmer Michael Phelps' unusually long arm span and oversized feet, it would be a little naive to pretend that Mäntyranta's unique genetic inheritance wasn't a factor in his path to success.

Owing to the vast diversity of human body sizes, wrestlers and boxers have long fought in weight classes. Stock car racers compete in a system in which all cars are built to roughly the same specifications. And, of course, men and women almost always compete separately in professional sports, since adult men tend to have a natural height, weight, and power advantage over adult women. All of that is a sometimes arbitrary way to keep the competition as fair as possible.

So is it inconceivable that we might one day compete in genetic classes, too?

Mäntyranta's turbocharged cardiovascular genetic inheritance, by the way, results from just a single letter change in his DNA. The change was in a gene that serves as the template for a protein that is the receptor for EPO. Instead of a G (for guanine) in nucleotide position 6002, Mäntyranta and about 30 members of his family have an A (for adenine) in a gene known as *EPOR*. This 0.00000003 percent change in Mäntyranta's genome was enough to cause the *EPOR* gene to make a protein that was really sensitive to EPO and that resulted in many more red blood cells being made. Yes, one letter in a field of billions is all it took for the corresponding protein made from the *EPOR* gene to give him a 50 percent increase in oxygen carrying capacity in his blood.[6]

We all carry these minor single letter or nucleotide changes in our genomes. The more related we are, the more similar our genomes. As we now know, since our genomes encode for templates that direct how our body is put together, the more similar your genome—think monozygotic or "identical" twins—the more you may physically look alike. Now, if you don't look at all like your siblings, that doesn't mean that you're not related. What's at work there is that you likely each inherited a different and unique combination of genes from your parents.

And what you've inherited has also been shaped by what all your ancestors have experienced. As we saw previously with lactose intolerance, if your ancestors didn't raise animals to consume their milk, then you're likely genetically out of luck when it comes to being able to enjoy ice cream into adulthood. And many of our adaptations don't end there.

Which brings us back to the Sherpas, who, given their unique genetic inheritance, have taken on—as a matter of cultural pride and economic necessity—the brunt of the dangerous burden of helping mountaineers from around the world reach the summit of the world's tallest mountain (which, at 29,029 feet, peaks just below the altitude at which most large commercial airliners fly). Among these amazing people is an unassuming man named Apa Sherpa, who as of 2013 jointly shared the world record for the most ascents of Everest, including four in which he climbed without the aid of supplemental oxygen. As a boy, Apa never intended to climb the mountain even once, but when he learned he was good at it, he discovered a way to help his family.[7]

How is it that he is so good at climbing to the top of a mountain that, until 1953, had never been touched by human feet? Indeed, how is it that Sherpas appear to be so well adapted to living in this high-altitude environment at all?

Well, as you may have guessed, some members of this ethnic community have inherited a very small genetic change that has led to profound differences in their lives. In their case, the change is in a gene called *EPAS1*. Rather than producing more red blood cells, this special Sherpa gene produces fewer, seemingly blunting Sherpas' biological response to EPO.

After everything I told you about the mighty Mäntyranta and his genetic inheritance, this might not seem to make sense, at first. After all, wouldn't Sherpas be better suited for their atmospheric existence if they were born with blood thick as honey that's brimming with red blood cells and thus loaded with oxygen?

Well, sure—for a while. But remember: While thick blood can be good for short periods of time, it can also be dangerous, increasing the odds of devastating strokes if allowed to endure for too long. The Sherpas don't just visit the Himalayan highlands, they live there. And so they don't need well-oxygenated blood just for skiing and cycling races, they need it all the time.

Instead of ever-rising levels of red blood cells under situations of decreased availability of oxygen, what the Sherpa's unique *EPAS1* genetic configuration provides them is stability over time—the ability to transmit adequate oxygen throughout the body even in conditions in which it's harder to come by in the surrounding atmosphere.

As unique genetic groups go, the Sherpas are really quite young. By way of context, their move toward Chomolungma is likely to have happened right around the time that Christopher Columbus was getting ready to sail to a place we'd ultimately come to call North America.

The Sherpa-specific *EPAS1* mutation might, in fact, be an example of natural selection at play—and some researchers believe it may be the fastest case of human evolution that has been documented so far.

In other words, the Sherpas' low-oxygen living conditions have quickly changed the genes they've inherited, which are now being passed down through generations.

And you've probably inherited such changes, too. Maybe not in your *EPOR* or your *EPAS1* genes, but likely in genes that helped your particular ancestors survive. As we map more genomes and become more and more familiar with the single nucleotide polymorphisms (changes in just one letter of a person's genetic code that we call SNPs) that are both subtly and magnificently diverse between groups of people around the world, the more light we'll shed on our ancestors' history—and the more, in turn, we'll discover about ourselves.

As I sat atop the peak of Mount Fuji and watched the sun slowly begin to crest over the early dawn sky, I couldn't believe how much my feet were killing me. Being so busy with the nausea and flatulence that accompanied my own ascent up the mountain, I had failed to notice that my feet had been left badly blistered and sore. After sitting quietly for a few minutes sipping my can of Coke, I slipped out of my boots to assess the damage. I imagined they didn't look as bad as they felt until I peeled back my socks all the way. My toes seemed to have borne the brunt of the climb. With all the rain my boots had become waterlogged, and this turned my toes into swollen and incredibly painful minisausages. And I knew what was coming: an hours-long descent down the mountain. As I thought about what to do next, I started fantasizing that besides being a little more genetically Sherpa so as to avoid altitude sickness, wouldn't it be nice to live a life completely free of pain?

AT SOME point in our lives we all become acquainted with some type of pain. It may be one of your earliest childhood memories. You may be feeling some right now. One thing's for sure: Pain, especially when it's of the chronic persuasion, is serious business. You may be surprised to learn that it's been estimated to cost up to $635 billion a year in the United States alone,[8] a figure greater than the costs associated with conditions like heart disease and cancer.

Staring at my toes atop Mount Fuji, I knew that the pain I was feeling was not serious and likely only temporary (at least I hoped so). Yet, unfortunately that's not the reality for millions of people whose lives are chronically debilitated by pain at a cost to which no dollar figure can be attributed.

While I contemplated putting my wet socks back onto my blistered feet, there was nothing I wanted more at that moment than at least a short reprieve from the throbbing ache. I imagined what it would be like to morph into some comic book character figure with superhuman abilities. And I knew I wasn't alone in my wishful fantasizing. Indeed, what most people wouldn't give to be impervious to pain. But before that wish gets granted, we need to meet a 12-year-old girl named Gabby Gingras.

Very quickly after she was born in 2001, Gabby's parents noticed that their infant girl was a bit unusual. She would scratch at her face. She jabbed her fingers in her eyes. She banged her head against her crib without crying. And when her teeth began to come in—an extremely painful experience for most children—Gabby didn't really seem to mind.[9]

Then there was the biting. Lots of children bite their parents and siblings. And teeth are, of course, a common reason why mothers stop nursing. But Gabby wasn't just biting other people. She was biting herself. She gnawed on her own tongue until it looked like raw hamburger. She chewed on her fingers until they were bloody.

It took months of doctors' visits to find the answer to why this beautiful baby girl was hurting herself: Gabby is one of a very small number of people around the world who suffers from a genetic condition called congenital insensitivity to pain with partial anhidrosis. This condition causes them to feel no pain in parts or in all of their body.

It's possible that more people are born with this very rare condition, but they don't survive very long—because, as it turns out, a life without pain is a life that is really hard to keep living.

Even after Gabby's parents understood why their daughter was injuring herself, there was little they could do to completely protect her. It would be years before Gabby was old enough to reason with. In the meantime, all they could do was try their best to protect her from herself. They made the difficult decision to preemptively pull all of her baby teeth from her mouth. That, however, led her adult teeth to grow in early—and those were promptly gone, too.

Though it was badly damaged by all of the poking she had done to it, doctors were able to save Gabby's right eye by sewing it shut for a time. Once it had healed as much as possible, Gabby was forced to wear swim goggles almost all the time. Her left eye, though, couldn't be saved; it was removed when she was three years old.

As much as we'd like not to think about it when it's there, pain actually protects us. It helps move us from infancy to maturity, and it provides the basic binary feedback we need to develop more advanced decision-making abilities. *It hurts when I touch this? Okay, I won't touch it anymore.*

For all of that to happen, though, your body must be able to transmit pain signals from one place to the next. Handing the message of pain from cell to cell and up to your brain like a microscopic Pony Express moving at electric speed is a process dependent on specific proteins.

This became apparent when mutations in a gene called *SCN9A* were discovered in a rare and related condition to that of Gabby's called congenital insensitivity to pain.[10] The difference between people who are insensitive to pain and others on this planet is just a small variation in the version of the *SCN9A* gene that we've inherited.

Changes in *SCN9A* and other related genes can lead to a family of diseases called channelopathies. The term simply refers to different conditions that are thought to result from nonfunctioning gates that sit on the surface of our cells and mediate, or determine, what goes in and what comes out. In the case of some people who feel no pain, the protein that is made from the *SCN9A* gene stops the signal from being sent. The message is dropped off, but instead of setting forth on a swift and wild adventure, the pony and its rider just dawdle at the corral.

The discovery of *SCN9A* and its association with pain transmission came after scientists from the Cambridge Institute for Medical Research decided to look into reports of a boy in Lahore, Pakistan, who was said to have a superhuman ability to be impervious to pain. As a human pincushion, he made a living from his apparent inability to feel pain in street performances, impaling himself with all sorts of sharp objects (none of them sterile), swallowing swords, walking on hot coals, and expressing not the slightest hint of being bothered by it all. He kept showing up at a local hospital to have wounds patched up after having stabbed himself with knives. By the time the scientists reached Lahore, tragically, the boy was dead, just shy of his fourteenth birthday, having jumped off a building in an attempt to impress his friends. Interviews with relatives in the boy's extended family revealed several others who reported that they'd never felt pain, and a dive into their genetic pool showed that all of them had one thing in common: the same mutation in their *SCN9A* gene. I'm always left awestruck by the incredible range of effects that emanate from the subtlest of changes in our genetic code and its expression. One single letter change in a series of billions of letters and you've got bones that break with the slightest of pressure. Another small shift in expression and you wouldn't feel that broken bone at all.

When it comes to pain, things have moved rather quickly since the discovery of the *SCN9A* gene. We now have a growing list of other genes (close to 400 already) that play an instrumental role in how our lives are impacted by pain. All of these discoveries are leading to a whole new line of research into how—in the very near future—we might hopefully be able to selectively dial down the intensity of some types of chronic pain. The *selective* part is key—because, as we learned from Gabby and the boy from Lahore, the protective effects of immediate pain are vital to our survival.

Many of the small differences in our genetic inheritance play a far greater role than mediating our response to pain. Figuring out how they all link together is the next big research challenge that I've become involved in unlocking.

WHEN THE human genome was first published, the rush was on to identify genes linked with specific traits—and most of the low-hanging fruit was picked clean pretty quickly. Many of the gene-linked conditions we've identified thus far are monogenic. As in the boy from Lahore who didn't feel pain, these changes can result from alterations in one single gene. Far trickier is the task of untangling the complicated web of factors that give rise to conditions such as diabetes and hypertension that likely involve more than one gene.

To get a sense of what that task is like, imagine trying to walk in a specific pattern from dormitory to classroom to courtyard to laboratory to library and back again across the unpredictably moving and shifting grand staircase at Harry Potter's Hogwarts School of Witchcraft and Wizardry. Just the slightest misstep and you're back where you started. That sort of complexity can be mind-boggling and frustrating, particularly when the stakes are, quite often, literally a matter of life and death.

Today, the movement in genetics is not only to look at specific genes and what they do but rather to better appreciate what our genetic inheritance does as a network—and, of course, to understand how our life experiences impact that intricate system through mechanisms like epigenetics.

Further complicating matters is the even more difficult challenge of understanding how the life experiences of our parents and other relatively recent predecessors are also impacting our current and varied genetic landscapes.

Knowing what these changes mean to us personally will help us make better decisions about everything from what kinds of adventures we should undertake (no more mountain climbing for me), where we live (you won't find me moving to Alma, Colorado, elevation 10,578 feet, any time soon), and, as we discussed in detail in chapter 5, what we eat (I still really love my semolina gnocchi, though I prefer to eat it at sea level).

All of those things—and so much more—that we have been genetically gifted with are part and parcel of our own unique inheritance.

Other than the Coke machine and my aching feet, I don't remember much else from being on the peak of Mount Fuji. But I do remember seeing the sunrise. And I remember looking around, in that moment, at the faces of all of the people who were sharing that experience with me. There were people of all ages. Some looked fresh and rejuvenated as though they'd spent the night soundly asleep and had not just climbed up a mountain—appearing bright as that morning sun—while others, like me, looked as though they were ready to collapse.

And soon after the sun broke through the clouds on the horizon, we were all on our way.

Our guide came over with an outstretched arm, pointing toward somewhere beneath the clouds. It was time for us to make our way back down the mountain. As I collected my pack, I fumbled around, searching for a fresh pair of socks to prepare for the descent. I couldn't help thinking that despite not having Sherpa genes, I managed to make it to the summit of Mount Fuji. Which for me was symbolic of the human capacity to surmount the supposed limitations of our genetic inheritance. After all, being a superhero has more to do with making superhero choices, day after day, irrespective of the genes we've inherited.

CHAPTER 9

Hacking Your Genome

Why Big Tobacco, Insurance Companies, Your Doctor, and Even Your Lover All Want to Decode Your DNA

Cancer is the black death of our time. And that, in and of itself, can be seen as something of a success. After all, we've come extremely far in taming many of the infectious diseases that were the top killers of our kin for most of human history. Today, in the developed world, one of the greatest dangers comes to us not from rats or ticks, viruses or bacteria, but rather from within us.

About 7.6 million people die of cancer every year across the globe. Fill a room with 10 people and you'll have four among that group who will be diagnosed with some form of cancer in their lifetime.[1] Do you know anyone whose family hasn't been touched, in some way, by this disease? I don't. And I don't know anyone who hasn't considered that they or someone they love might one day get it, too.

This isn't a new curse. Some anthropological archaeologists believe Egypt's longest-reigning female pharaoh, Hatshepsut, may have died from complications related to cancer.[2] Going back even deeper into our evolutionary history, paleontologists have found fossilized skeletal evidence that dinosaurs—in particular the duck-billed hadrosaurs (Late Cretaceous herbivores who were known to eat the leaves and cones of what we think of as carcinogenic conifer trees)—suffered this fate as well.[3]

Among our own species these days, the most prolific of these malignant killers is lung cancer.[4] But while we know that 80 to 90 percent of lung cancer cases involve people who smoke, we also know that not everyone who smokes is equally likely to develop lung cancer.[5]

Take George Burns. In one of his final interviews the then 98-year-old comedian told *Cigar Aficionado* magazine, "If I'd taken my doctor's advice and quit smoking when he advised me to, I wouldn't have lived to go to his funeral."[6] Did Burns' penchant for cigars—10 to 15 of them a day over the course of 70 years—contribute to his longevity? Unlikely. But as far as we can tell, all those El Productos don't seem to have shortened his life, either.

Some people wrongly mistake cases like this as evidence against the prevailing wisdom that tobacco is bad for you. It's not evidence of that at all. But it is fair to say that just because a habit—be it compulsive smoking, drinking, or binge eating—makes adverse health effects *more* likely (people who smoke are 15 to 30 times more likely to get lung cancer than nonsmokers, according to the Centers for Disease Control and Prevention), that's not the same thing as making it likely (only about one in 10 smokers will actually get lung cancer).

To be clear, though, smoking is Russian roulette. Not to mention expensive. And secondhand and thirdhand smoke puts other people—generally the ones we are closest to—at greater risk.

So why can some people smoke their entire lives and not get lung cancer? We haven't yet found the magic combination of genetic, epigenetic, behavioral, and environmental factors that can accurately predict who is at greatest risk. Pulling apart that tangled web is not going to be an easy task. But it's probable that a certain mix of genetic and environmental factors, might in fact play a role in decreasing your chances of developing lung cancer as a result of smoking. There hasn't been a lot of serious scientific work done in this area of human health. Not too many scientists out there are pining for an opportunity to do research that might have the perverse effect of telling certain groups of people that they don't have as much to worry about when they stick a cigarette between their lips.

There is one industry, though, that is keenly interested in this line of scientific exploration. And that's Big Tobacco.

HONEST SCIENTISTS have known since the 1920s about a probable link between smoking and lung cancer. And really, anyone who thought about it much at all could have reasonably concluded that sticking a burning piece of paper soaked with chemicals and stuffed with tobacco leaves, accelerants, insecticides, and who knows what else into your mouth isn't likely to be the panacea that cigarette companies sometimes claimed it was.

Yet, the health dangers went largely ignored by the public for the next three decades.

Then along came Roy Norr. When the veteran New York writer first published his medical exposé on the dangers of smoking in the October 1952 edition of the relatively obscure *Christian Herald* magazine, it didn't get much attention. But when *Reader's Digest*, then the most widely circulated magazine in the world, ran a condensed version of the same article a few months later, it was as though the floodgates had been sprung open.[7] Over the next few years, American newspapers and magazines published a barrage of damning articles linking tobacco use to "bronchiogenic carcinoma," as lung cancer was called back then.[8]

The reports were boosted by the increasingly sophisticated and quantifiable nature of scientific investigation that was being applied to medicine and that we take for granted today but which, in the 1950s, was still a relative rarity. We can consider this kind of research to be a success of science, but it was really born of a failure of humanity: A half century of world war, including the first uses of nuclear weapons, carpet bombings, and modern chemical and biological warfare, had made us experts at meting out and analyzing death. The sudden salvo against smoking was one of the first instances in which we really began beating all those quantitative swords into medical plowshares. It also came at a perfect time historically, as there was a concurrent wave of unprecedented funding for medical research following World War II.

But Big Tobacco was quick to strike back. At the time, more than 40 percent of American adults were regular smokers, and the average American smoker was lighting up 10,500 sticks a year. That was roughly a whopping 500 billion cigarettes annually.[9]

Big Tobacco was making a killing. And it wasn't alone. Back then, every time a pack of cigarettes was sold, the U.S. government picked up a cool seven cents.[10] Over a year, that amounted to $1.5 billion—the equivalent of about $13 billion today. That's not to mention all the jobs smokers were supporting in tobacco legacy states like Virginia, Kentucky, and North Carolina.[11]

Against the flood of bad press, Big Tobacco had to look like it was doing *something*. So, in what they called "A Frank Statement to Cigarette Smokers," the heads of 14 tobacco companies came together to publish a full-page advertisement in more than 400 newspapers nationwide. In it, they made the audacious argument that recent studies linking smoking to disease were "not regarded as conclusive in the field of cancer research."

"We believe that the products we make are not injurious to health," the tobacco chiefs' statement continued. "For more than 300 years tobacco has given solace, relaxation, and enjoyment to mankind. At one time or another during those years critics have held it responsible for practically every disease of the human body. One by one these charges have been abandoned for lack of evidence."

But in the same ad—and in spite of their public stance of incredulity—the collective heads of Big Tobacco pledged to do something rather remarkable. They would create the Tobacco Institute Research Committee, an independent scientific body of inquiry that would be responsible for reviewing the latest studies and conducting research of its own to best understand the health implications of smoking.

Perhaps not surprisingly, though, the committee (later renamed the Council for Tobacco Research) wasn't really independent at all—and its *real* mission was downright diabolical. Over the next few decades, the organization's researchers collected thousands of scientific papers and press clippings, looking for inconsistencies and instances of contrary results. It then used that information to formulate carefully crafted marketing messages, fight legal actions and regulation, and continue to sow doubt as to the real dangers of smoking.

Leading this mission of misinformation was Clarence Cook Little, a geneticist whose academic work on Mendelian inheritance had been extremely influential in the years before World War I and whose wide-ranging curriculum vitae included stints as president of the University of Maine and University of Michigan as well as more controversial roles as president of both the American Birth Control League and the American Eugenics Society.

But the line on Little's résumé that the tobacco companies really coveted was his tenure as managing director of the American Society for the Control of Cancer, the forerunner of today's American Cancer Society.

Appearing as a guest on Edward R. Murrow's television show *See It Now* in 1955, Little was asked if any cancer-causing agents had been identified in cigarettes.

"No," he answered. He then said, in a thick New England accent, "None whatever, either in cigarettes or any products of smoking, as such."[12]

It wasn't supposed to be a punch line, but over the past half century that TV segment (which includes Little chewing on the end of what appears to be an unlit pipe) has been played over and over to great comedic effect.

To Little's Teflon credit, though, his full answer was a little more nuanced. "This is interesting in a way," he continued, "because there are many known cancer-forming substances in tar, and I'm sure that research in this field will continue. People are bound to look for cancer-causing agents in all kinds of material."

So cigarettes don't cause cancer, but tar from smoking them—which invariably builds up in the lungs—does? If Little wasn't already sitting so comfortably in the pocket of the tobacco companies, he might have had a second career as a politician. As George Orwell said, such artful dodges are "designed to make lies sound truthful and murder respectable."

While Little might have danced around the truth, though, he wasn't lying. Not strictly speaking, anyway. Because, after all, most of the research being done at the time was looking for a direct and specific association between the immediate act of smoking and lung cancer, and the sophisticated tools for homing in on what was actually causing cells to turn from friendly to malignant was still many years away.

But for our purposes, something else that Little said on that evening is even more interesting—something that may be a clue to what's to come, not just from the tobacco industry but from anyone who has produced a product that can make people sick.

"We're very interested," he went on to say, "in finding out what kind of people are heavy smokers and what kind are not. Not everybody is a smoker. Not everybody who smokes is an equally heavy smoker. What determines these selections on the part of people? Is it a different nervous type of person who smokes a great deal? Is it a person who is reacting differently to strain or stress? Because it is very clear that certain people just can't take it as well as others."

Very interested? Of course Big Tobacco would be. And of course it still is. If the tobacco industry can establish why certain people are more likely to be heavy smokers—and thus more likely to become sick—then it can shift the blame, arguing that it's an inherited and possibly genetic susceptibility to heavy smoking that is the problem and not the cigarettes themselves.

If you haven't already heard the same type of talk from soft drink and junk food manufacturers, just keep your ears open. It's coming. And the next time someone sues a fast food chain for making them fat (as one McDonald's manager did in Brazil a few years back) you can be certain that the plaintiff's genome (and bacterial microbiome, too) will likely be on the defendant's expert witness list.

Because when it comes to absolution from responsibility, big business has a history of, as *The Godfather*'s Sonny Corleone might say, "We go to the mattresses..."

Want proof? Look no further than BNSF, the Burlington Northern Santa Fe railroad.

OUR BODIES weren't meant to behave in this way.

We're active animals. Or we once were. Our prehistoric days used to be just a little more physically lively. Pouncing on small game, climbing over rocks, swimming across rivers, and running from saber-toothed cats.[13]

But ever since the Industrial Revolution—and especially since the

Digital one—two big changes have happened: We've become sedentary, and our lives exceedingly repetitive.

Only in the past few centuries have we subjected our bodies to the types of physical beatings that come with doing the same things thousands, even millions, of times over. And from carpal tunnel syndrome to lower back pain, our joints and bodies are paying the price.

We owe our understanding of repetitive strain injuries to the father of occupational therapy, an Italian physician by the name of Bernardino Ramazzini. His book *De Morbis Artificum Diatriba* or *Diseases of Workers* was published in Modena, Italy, in 1700 and is still quoted by those who work in public health.

What could a seventeenth-century Italian doctor possibly have to say about twenty-first-century office life? Well let's take a look at *De Morbis* to find out:

> *The maladies that afflict the clerks…arise from three causes: First, constant sitting, secondly the incessant movement of the hand and always in the same direction, thirdly the strain on the mind from the effort not to disfigure the books by errors or cause loss to their employers when they add, subtract, or do other sums in arithmetic…Incessant driving of the pen over paper causes intense fatigue of the hand and the whole arm because of the continuous and almost tonic strain on the muscles and tendons, which in course of time results in failure of power in the right hand…* [14]

Yes, he pretty much nailed it, describing succinctly what we call today repetitive strain injuries.

What Ramazzini recognized more than 300 years ago was that the process of doing the same thing over and over again is simply bad for us.

And that brings us to the BNSF railroad. The company was

founded in 1849 in the Midwestern United States and today has grown into one of the largest freight railways in North America, with lines cutting across 28 states and two Canadian provinces.

It takes almost 40,000 workers to keep all those trains on the track. And as you can imagine, working on a railroad can be physically tough. That's why it's not surprising to find out that some of BNSF's employees were occasionally taking disability leave because of work-related injuries. This, of course, can be extremely expensive for employers like BNSF, which prompted their management team to explore ways to keep costs down.

Now, one good way to do this would have been to become even more vigilant about improving occupational health standards. They didn't do that. Another way would be to ensure all workers are encouraged to take even more frequent breaks or rotate off from repetitive and injury-provoking activities. They didn't do that either.

Instead, they went after the genes of their employees.[15]

You see, someone in BNSF management had apparently become interested in genetics upon learning that DNA might play a key role in determining whether a person is susceptible to the symptoms of tingling, weakness, and pain in the hands and fingers that we've come to identify as carpal tunnel syndrome.[16] Soon, according to the U.S. Equal Employment Opportunity Commission, BNSF employees who filed claims for work-related injuries based on carpal tunnel were being forced to submit to having their blood drawn. The blood was then—without the employee's knowledge or consent—allegedly tested for a DNA marker that would show whether the employee was genetically susceptible to wrist pain and injury.

Purportedly, faced with the prospect of losing their jobs if they refused to be tested, most of the workers permitted their blood to be drawn. But at least one employee decided to fight back. Eventually there was a $2.2 million dollar settlement between BNSF and the

Equal Employment Opportunity Commission, which took up the cause of the employees on the grounds that the testing violated the Americans with Disabilities Act.

That was in the early 2000s. Today, U.S. federal law protects individuals from genetic discrimination in the workplace. The Genetic Information and Nondiscrimination Act, or GINA, was created to protect people from genetic discrimination in situations pertaining to employment and health insurance. Signed into law by President George W. Bush in 2008, the legislation that some called "the anti-Gattaca law" (rumor has it that some politicians were moved to support the measure after seeing the 1997 film about a genetically tiered future society) was heralded as a significant step forward in trying to predict and prevent some of the discrimination people might face as a result of genetic testing.

Unfortunately, though, GINA provides no protection against discrimination in matters of life and disability insurance. This means that if you've inherited a genetic mutation, say in your *BRCA1* gene, that could play a role in reducing your life expectancy or could make you more susceptible to disability, your insurance company can legally charge you more or outright deny you this type of coverage. This is why I always ask my patients to consider carefully what they might be getting themselves and their families into before they get any genetic testing or sequencing done that is not done anonymously. Because what we discover—while potentially vital for your health—can also become a disqualifying factor for disability and life insurance for you personally, your immediate family, and all your future genetic descendants.

As genetic testing and sequencing becomes more routinely used in different aspects of medical care, from pediatrics to gerontology, we'll have more information at our disposal to be able to link distinct health risks with our unique genetic inheritance.

Obamacare is set to give many Americans better access to health care, but it might also inadvertently set them up for genetic discrimination. Thanks to a glaring loophole intentionally crafted into GINA, insurance companies have free reign to use that genetic information against us when they determine the premiums they will charge us for disability and life insurance.

Here's where things get even more frightening. These days, a potential insurance provider, or anyone else for that matter, doesn't have to touch a single one of your cells to get a lot of information about your genetic inheritance.

Among scientists like myself, it's common practice to share genetic and sequencing data with other researchers while removing identifying information such as names and social security numbers. But what most of us have always seen as a relatively solid privacy protocol, an astute team of biomedical experts, ethicists, and computer scientists from Harvard, MIT, Baylor, and Tel Aviv University saw as a potential target to be hacked.

Plugging short segments of seemingly anonymous information into recreational genealogy websites (the users of which are increasingly including genetic information as a way to track down long-lost family members), the researchers were able to easily identify the anonymous patients' family groups. And with just a little additional data that's commonly included in shared samples—age and state of residence, for instance—they were able to triangulate the precise identity of many individuals.[17]

This can also work the other way around. Do you have a family member who has survived cancer? Did they keep a blog? Facebook it? Tweet about it? Social media's not just a great way to keep in touch with our loved ones—it's also a potentially very deep and rich source of information for genetic cyber sleuths. Already, more than a third of employers say they've used information found on social media sites such as Facebook to eliminate job seekers from the applicant pool.[18] With employer-based health-care costs in the United States rising ever skyward, companies might feel justified in making a social media health status sweep a regular, if secretive, part of their hiring practices.

Using just your name and the millions of genealogy records publicly available on the Web, an inquisitive and resourceful person—someone considering hiring, dating, or marrying you, perhaps—could come to know more about you than you might even know about yourself.[19] And if you just happen to be that resourceful and inquisitive person and there was a much easier way for you to access someone's genetic information without them ever knowing, how far would you go? What I'm asking you is this: Would you be willing to hack someone's genome?

I WAS trying to hail a cab when my phone vibrated to let me know that a new e-mail had arrived. It was from a friend of mine, a young professional named David who was recently engaged. His fiancée, Lisa, was a fashion photographer who was also living in New York City. I had the pleasure of meeting her just a few weeks prior to their official engagement, at her first solo photography exhibit in a gallery in SoHo.

David e-mailed that night to ask if I was free to chat as he had a few questions he wanted to ask me about genetic testing. This is a common request I get from friends and family who are looking for advice in this rapidly evolving field. David had mentioned that he was looking forward to starting a family with Lisa once they got married, and I assumed that he wanted to take advantage of expanded prenatal genetic testing possibilities. These "gene panels" can be used to see if you and your partner carry mutations in hundreds of genes. This type of testing can provide couples with a genetic snapshot of their genetic compatibilities. We all carry around a handful of recessive mutations. On their own they are mostly harmless, but if you and your partner share the same misbehaving gene, that's a recipe for potential parental genetic disaster. Many more couples are taking advantage of screening hundreds of genes before starting down the road to parenthood. And it's easy to do: Just spit into a small vial, drop it into the mail, and wait for the results.

Given the fact that most of our mutations aren't in the same genes as our partners', though, this type of genetic incompatibility is often avoided. But as I quickly came to discover, after finally getting into an available cab and giving David a call, prenatal testing was not what he was after. Instead, he wanted to know whether he could hack his fiancée's genome without her knowledge.

David's concerns arose after his fiancée, who was adopted when she was very young, was reunited with her biological father. Lisa had tracked her father down in anticipation of inviting him to their wedding. A coffeehouse conversation had revealed that her biological mother had passed away after suffering from a slew of symptoms that sounded a lot like Huntington's disease, a fatal genetically inherited neurodegenerative disorder.

In someone who has Huntington's disease the nerve cells in the brain slowly degenerate. There's no cure for Huntington's, and the path to death is paved with a loss of muscle coordination, psychiatric problems, and eventually cognitive decline and death.

Complicating matters, though, was the fact that David's fiancée wasn't interested in getting genetically tested herself.

"But," he said, "if I could just get you a piece of her hair, or something like her toothbrush, that's all it would take, right? We could check, right? I mean, she wouldn't even have to know. I understand this is crazy. But...it would be so much easier if I knew what I was up against."

What he was asking me to facilitate was ethically problematic at best and, in a lot of countries, downright illegal.[20] Rather than expressing my complete disapproval outright and declining his request, leaving him to his own devices, I thought it best to invite him out for a drink. David said he needed to run a few errands after work and would be free later. We agreed to meet at 10 p.m. I was looking forward to uncovering what would have led David to consider behaving so uncharacteristically.

It was one of those exasperatingly hot and humid August evenings in Manhattan when most people seek shelter in an air-conditioned dwelling or, if they can, leave the city. As I got out of my cab and ducked into the bar, I was really glad to get a reprieve from the humidity.

I found two free stools at the bar, sat down and ordered. Watching the bartender expertly prepare and pour my muddled mojito, I thought about David and decided to call Kelly. She's a social worker and friend, who has a lot of experience counseling and working with the partners of people newly diagnosed with a terminal illness.

"Try to identify some of the underlying fears and expectations associated with getting married to someone who might be harboring the gene for a fatal inherited condition," Kelly said. "Then figure out what kind of discussions they've already had. Most of us are scared to be vulnerable—especially in front of our partners—but if he hasn't expressed his fears to her, neither of them can have an honest conversation about what this means for their future, their relationship, and what to do next."

A few minutes later, David walked into the bar. Not surprisingly, he wasn't interested in having a conversation about applied medical ethics. What he wanted was simply to be heard.

As the evening progressed, I was reminded that *not knowing* is sometimes much more complicated and painful than knowing. Having been friends with David for many years, it was obvious to me that he was experiencing a lot of emotional pain, not to mention shock. He felt that the person he wanted to spend his life with was keeping a secret inside of her that she didn't want to let out.

I tried my best to just sit, listen, and answer only the questions that I actually had answers for, which, to be honest, weren't many. As the night progressed, I heard about the surprise discovery that Lisa's biological father was alive and living not too far from them in up- state New York. I heard of the painful revelation that her mother had passed away young, leaving so many unanswered questions. I heard about the frustration David felt toward Lisa's seeming ambivalence and staunch resistance to being tested.

"I don't get why she doesn't want to know," he kept saying.

In this digital age, David already knew a lot about Huntington's disease. He'd learned that, unlike other conditions that are caused by specific single letter mutations, the genetics behind Huntington's dis- ease could be compared to a scratched record that keeps skipping. People with this devastating neurological condition have, in a gene called *HTT*, a longer than normal stretch of three nucleotides— cytosine, adenine, and guanine—that repeat themselves over and over again.

We all inherit a certain number of these repeats, but when some- one has a gene that has 40 or more, they almost always develop Huntington's disease. The more repeats, the earlier the disease de- velops. If there are more than 60 repeats, the affected person might develop symptoms of Huntington's as early as the age of two.

It's not exactly clear why, but most people who develop Hunting- ton's very young have acquired the gene from their father. But even in the case of those who get it from their mothers, the repeats gen- erally increase from one generation to the next. We call this type of change in genetic inheritance *anticipation*.

From our conversation, it seemed that David had a pretty good handle on all of this material, including the way in which the disease was passed down. And because all you need is one copy of the *HTT* gene with a greater than normal number of repeats, he knew that if Lisa's mother was affected, Lisa had a 50 percent chance of inheriting Huntington's disease. And if that was the case, given the mechanism of anticipation, she would likely start showing symptoms at a younger age than her mother did when she first got sick.

And most of all, he knew that, if she did have it, he wasn't going to get to grow old with her. Instead, he would have to watch her personality change as the disease remodeled her brain, slowly disassembling her mind. Would he have the emotional, mental, and physical strength to properly care for her needs?

"But I can do this," he said. "Look, I know testing her for Huntington's without her consent is wrong. But I just wish I knew what we're up against. It's the *not* knowing that's killing me. Why can't she just get herself tested? Maybe having an answer either way would make us live our lives differently... but ultimately, I guess, the choice to test is hers to make."

And that was it. David ended the conversation abruptly. I asked for the check and got ready to face a hot and sticky cab ride home.

I'd really love to tell you that there's a happy ending to this story.

I wish I could say that they're living a fabulous life together in a trendy neighborhood in Brooklyn, just like they'd planned. And that David found the strength to approach Lisa one more time and she agreed to undergo testing.

And more than anything, I'd love to tell you that Lisa tested negative for Huntington's disease.

But genetic stories are like the rest of life. Sometimes they're incredibly beautiful and other times they're terribly painful. And sometimes they're somewhere in between.

The truth is that David and Lisa didn't get married as planned. She's still wearing his ring, and they're still madly in love—maddening as love and life can sometimes be. For his part, David is still trying to come to terms with Lisa's reluctance and resistance to knowing what lies ahead for both of them. For her part, Lisa has been in touch with a counselor who specializes in helping families affected by Huntington's disease, though as of this writing, she hadn't made any decisions about whether to be screened.

As the cost of genetic testing continues to plummet and as it continues to get easier to do, we'll face more of these situations—and for many more conditions. To hack or not to hack a genome is going to be the question we'll increasingly be faced with. And we're not always going to have the ethical sophistication and experience to deal with the implications of that question.

As we enter more fully into this brave new world, our relationships will be tested, and our lives will change. And as we're about to see, so will our bodies.

ANGELINA JOLIE knew her odds were not good.

The Academy Award–winning actress had watched—feeling helpless despite all her status and fame—as her mother lost a years-long battle with cancer. Wanting to ensure she would be able to continue being there for her own partner and their children, she underwent genetic testing that revealed a mutation in her *BRCA1* gene.

In most women, a *BRCA1* mutation can mean about a 65 percent chance of developing breast cancer. That's because *BRCA1* belongs to a group of genes that, when functional, suppress tumors from forming by dialing down any rapid or uncalled-for growth.

But that's not all the *BRCA1* gene can do. It can also work in concert with a slew of other genes to repair damaged DNA.

So far, we've talked a lot about how many of our behaviors can change the expression of our genes through mechanisms like epigenetics. What you may not be aware of, though, is that many of the things you do on a daily basis can actually physically damage your DNA. And unbeknownst to you, you've likely been abusing your genome for years.

In fact, if there were a governmental agency called the Department of Genetic Protective Services, it would have removed your genes long ago to protect them from you.

Even something as seemingly positive as a short, relaxing vacation abroad can be surprisingly bad for you. And your rap sheet would probably look something like this:

1. Air travel between the United States and the Caribbean—check.
2. Stayed out in the sun too long working on your tan—check.
3. Consumption of two poolside daiquiris—check.
4. Secondhand tobacco smoke—check.
5. Exposure to insecticides, used for bedbug control—check.
6. Nonoxynol-9, found in contraceptive lubricant—check.

I'm sorry to have to ruin your recent fictitious holiday vacation in this way. But the Department of Genetic Protective Services is leveling these charges against you to try to make you appreciate just how much you take your genome for granted.

Everything on that list can damage your DNA. Without the ability to continuously and properly repair all of the negative changes we cause to our genome, we would run into some serious trouble. How good we are at repairing genetic damage has a lot to do with the "repair" genes we've inherited. If you happened to have inherited one of the more than a thousand known mutations in the *BRCA1* gene that can predispose you to cancer, then you need to be extra careful with the way you treat your genes. And yet interestingly, not all of these inherited mutations are equally worrisome.

Which brings us back to Angelina Jolie. When they tested her *BRCA1* gene, her doctors told her that her particular genetic variant or mutation was not reassuring at all.[21] There was, they said, an 87 percent chance that she would develop breast cancer and a 50 percent risk that she would get ovarian cancer.

Over a three-month period in the winter and spring of 2013, one of the most closely watched women in the world emulated some of her on-screen espionage-savvy characters, managing to elude the paparazzi as she underwent a series of procedures at the Pink Lotus Breast Center in Beverly Hills, California, including a double mastectomy.[22]

"You wake up with drain tubes and expanders in your breasts," Jolie wrote in the *New York Times* shortly after the procedure was completed. "It does feel like a scene out of a science-fiction film."

And not too long ago, it would have been.

Doctors have been performing mastectomies for a long time, but until quite recently it was a surgery intended to remove disease, not to prevent it.

That all changed, though, as our understanding of the molecular underpinnings of cancer matured and genetic screening and testing became more available and subsequently more women (and even some men) began getting the terrifying news that Jolie received. Faced with the decision to undergo a significant but imperfect screening regimen, about a third of these women are now opting for a preventative mastectomy. Having their breasts preemptively removed before cancer could strike. In doing so, they've created an entire new class of patient: the previvor.

The previvors are already thousands strong—almost entirely women who have had to face the same decisions as Jolie. As we come to better understand the genetic factors at play in other diseases—colon, thyroid, stomach, and pancreatic cancers are among the likely contenders—it's almost certain that this group of people will get bigger still.

"Cancer is still a word that strikes fear into people's hearts, producing a deep sense of powerlessness," Jolie wrote. But today, she noted, a simple test can help people understand if they are highly susceptible, "and then take action."

All of that is creating a whole new set of ethical complications for physicians, who foremost practice by the dictum *primum non nocere*.*

When it comes to action, we're not just talking about radical surgeries like mastectomies, colectomies, and gastrectomies. Because, of course, some things you just cannot remove. So other preemptive actions that people could employ will include increased surveillance or screening, preventative drug regimens, and, where it's possible, avoiding potentially damaging genetic triggers.

Which is why that rap sheet may turn out to be an important reminder of all the things you can do to take care of your genetic inheritance. If you don't take care of your genes, you may inadvertently change them.

Exposure to radiation during routine air travel, ultraviolet radiation while working on your tan, ethanol in your cocktail, chemical residues in tobacco smoke, insecticides, and chemicals in your personal care products, are all examples of general factors that can damage your DNA. How you choose to live will determine how well you treat your genome.

This means we all need to be better educated, not just by uncovering our family's medical history and by decoding our own genetic inheritance but by investigating what proactive and positive changes

---

* Latin, meaning "First do no harm."

we can make in our life with that information. These proactive changes will require different actions from each of us. For some of us that means avoiding fruit, while for others it could mean a mastectomy.

At the same time we also need to appreciate how others might use this information in an accelerated genetic future. And "others," as we've already seen, will include your doctors, insurance companies, corporations, government agencies, and very likely your loved ones as well. Even though we might have expectations of confidentiality, we also need to keep in mind the real lack of protection against discrimination for life and disability insurance before considering hacking our own genome.

We're not just standing at the precipice of a tremendous paradigm shift; a lot of us have already gone over the edge. And because we are so connected, technologically and genetically, a lot more of us will be going over as well, whether we like it or not.

CHAPTER 10

Mail-Order Child

The Unintended Consequences of Submarines, Sonar, and Duplicated Genes

It started out as a quiet morning in the Caribbean. It was Thursday, May 13, 1943, and the SS *Nickeliner*, an American merchant ship uniquely configured to carry large quantities of ammonia, was carrying a 3,400-ton stockpile of the volatile cargo that was ultimately destined for England. Ammonia was an essential ingredient to make munitions that was in short supply during the war, and to get it to England necessitated a perilous journey across the ocean during the climaxing months of the Battle of the Atlantic during World War II.[1]

As for the *Nickeliner*'s 31-man crew, the day ahead would be anything but routine. That's because a German submarine, captained by a 35-year-old naval officer named Reiner Dierksen, had been tracking the ship from the moment she'd left port.

Six miles north of Manati, Cuba, a steel periscope belonging to the German submarine quietly broke the surface of the water. Slowly, deliberately, Dierksen's torpedo men lined up their shot. With his target confirmed, the veteran captain—who already had been responsible for sinking 10 Allied ships—gave the order to fire. Two German torpedoes entered the water, propellers whirling, gaining speed. There was a tremendous explosion—water and fire shooting a hundred feet into the sky. Soon the *Nickeliner* was at the bottom of the sea, her crew left to their fate in life rafts.

For the Allies, the problem was both simple and maddeningly complex: They needed a way to locate the submarines once they were submerged.

They found their answer in sonar. At the time, it was all in caps—SONAR—an acronym for *so*und *na*vigation and *r*anging. A large amplifier would produce an underwater *ping*, and a receiver would "listen" for the sounds to bounce back, which could then be used to roughly derive the distance to their target.

Seventy years later, navies around the world still use sonar technology as a key part of their countersubmarine and antimining efforts. But over the years we've found that's not all sonar is good for. Today, a technology that was originally designed to take life from this world has become a mainstay of those who help bring it in.

As thousands of sonar men returned home from the war in the late 1940s, they began experimenting with other uses for the technology. Some of the earliest adopters were gynecologists, who quickly learned that medical sonar, as it was originally called, could be used to detect gynecological tumors and other growths without invasive exploratory surgeries.[2]

Where sonar really came into vogue, though, was when obstetricians learned they could use it to see images of a fetus and its placenta, starting just a few weeks after implantation. Giving them what must, at the time, have seemed like a magical ability to watch firsthand as the baby's developmental stages unfolded. What most people don't realize even today is that these images also convey the very delicate genetic interplay between the expression and repression of genes during fetal life that play a significantly important role in our development.

Fetal ultrasounds, as they are known today, allowed physicians for the very first time to catch an early glimpse of any genetic missteps or abnormalities that previously lay hidden until delivery.

Before we move on to learn about the influence of genetics in our development, let's go back in time for a moment to answer the question: Whatever happened to the German submarine in WWII that attacked and sank the *Nickeliner*?

Two days after the sinking, a U.S. patrol plane spotted what appeared to be a surfacing U-boat. The plane released a marker into the water to indicate its position. As the German crew worked desperately to dive their sub back into the relative safety of the deep, an Allied ship sped out to the location where the sub had been spotted and, using its newly minted sonar device, was able to locate the submarine under the water.

Using the depth and direction information provided by the sonar device, the patrol plane's crew dropped three depth charges into the water. And with that, like a shredded aluminum can, the Nazi sub joined the *Nickeliner* on the bottom of the ocean.[3]

What started out as SONAR technology to find hidden submarines has today, without a doubt, become inestimably important in helping bring babies into the world. The one thing that no one could have ever imagined was that a technology that was initially developed to take life could return, so quickly after its reprieve, to once again selectively take it.

A technology that we've developed for one purpose often becomes repurposed in surprising ways. As you can imagine, in countries where more cultural currency is given to male children than female ones, the use of ultrasounds has become extremely problematic. When the value of gender is asymmetrical, the ability to tell a baby's sex before delivery allows parents to choose their child's gender.

Which is exactly what has happened in China. For many years, China has enforced strict and sometimes obligatory population control policies that limit most parents to having one child. The cultural importance of having a son in China, combined with the one child policy, has created an even greater pressure for pregnant parents to have a boy. The results—an excess of 30 million Chinese males, an imbalance created by the use of ultrasounds to systematically find and abort female pregnancies—speak for themselves.[4] And the practice is thought to be spreading.

Indeed, researchers have shown that when ultrasound technology has made its way to areas of China where it hasn't been before, the imbalance between male and female births increases.[5]

Ultrasounds also helped spark another trend, rather benign in comparison, which is still raging today. One it's likely you're guilty of participating in and supporting as well.

The advent of gender-specific clothing for babies and toddlers in the United States really began to take shape in the postwar period, and calcified as prenatal ultrasounds became more widely available across the U.S.—friends, family members, and colleagues simply had more time to go shopping, and the gender-specific baby shower was born.[6]

But where some see pink and blue, trucks and kittens, camouflage and lace, I see the cultural effects of what was, in effect, the world's first widely available prenatal genetic test. After all, for the greater part of the past century we generally agreed that, on a chromosomal level, the major difference between females and males is that the latter has a Y chromosome while the former does not. More than a fuzzy picture of our babies-to-be, the advent of prenatal ultrasounds provides us with a snapshot of the DNA they've inherited.

While ultrasounds can give us quite precise anatomical information, such as gender, generally by the fourth month of pregnancy, in the modern world of in vitro fertilization and preimplantation sex selection, we don't have to wait to find out. Which is why, if emerging and increasingly available medical technologies are not coupled with social and educational initiatives aimed at honoring girls in the same way as boys, things could get even worse.

And, of course, the amount of information we can now derive from basic genetic tests before pregnancy, or quite early into gestation, can tell us far more than simply gender.

Which, I suppose, might suggest that gender is a simple thing.

It's not.

A BOY or a girl? That's usually your first question when you learn that someone has had a baby, isn't it? And most of the time, that question appears to have a simple binary answer.

Gender identity is dependent upon a veritable rainbow of influencers, but when a baby first emerges from its mother's womb all that's really visible is the external plumbing. As one precocious five-year-old explained to Arnold Schwarzenegger's character in *Kindergarten Cop*, "Boys have a penis. Girls have a vagina."

The thing is, though, that's not always the case. Today we use the term *disorders of sex development*, or DSD, to refer to children and adults whose bodies have taken an alternative route along the pathways involved in the development of their reproductive organs.

Some of these paths can result in a significant amount of ambiguity in their outer genitalia—for instance, an unusually enlarged clitoris that appears to be a penis and labial lips that are fused and look somewhat like a scrotum. For physicians, it can be hard enough to keep up with the ever-changing spectrum of psychosocial understandings of sexuality. Likewise, we are now learning that the development of our physical sex mirrors that wide spectrum. This has left the classically basic and narrow "XY-means-male and XX-means-female" model of sex largely out of date.

In a world in which gender is still tied to everything from given names, pronouns, clothing styles, and public washroom segregation, ambiguity can cause a significant amount of embarrassment and consternation, particularly when there's uncertainty surrounding a baby's sex.

That's why, rather than just being a slight parental concern, gender ambiguity is often treated as a medical emergency—one for which doctors like me are called upon for consultations at all hours of the day and night.

So let me walk you through what happens when a child is born who is thought to have a DSD. Given the depth of the psychosocial issues at hand, we usually drop whatever nonemergency work we're doing and head over to meet with the family and medical team caring for these precious little patients.

Immediately thereafter, we try to get as much information as possible from the parents about their newborn child's family tree, including siblings, nieces, nephews, aunts, uncles, grandparents, and as many people up and down the line as possible. During this process we ask lots of questions. Are living relatives healthy? Is there a history of recurrent miscarriages or children with severe learning disabilities? Are the parents or grandparents or great-grandparents related in some way?

These questions don't just give us valuable genetic information, they also help remind everyone involved that the young baby is rooted in and part of a larger family—and most importantly, is not just a medical *problem* that needs solving.

We then move on to a physical examination that begins with the same sort of dysmorphology assessment we went through together in chapter 1, but in a lot more detail. With a hospital-issued measuring tape dangling from our necks and darting between our fingers, we check the circumference of the baby's head, the distance between the eyes, the distance between the pupils, the length of the philtrum, and so forth. We measure the arms, legs, hands, and feet. We also measure the length of the clitoris and penis and check to see that the anus is properly placed. Even something like the distance between a baby's nipples can occasionally give us valuable information about what's going on inside the infant's genome. Most importantly, when assessing for a DSD, we try to determine whether a baby appears to be dysmorphic overall.

It's not uncommon for people who are watching us perform these examinations to joke that we look more like tailors taking measurements for custom-made baby clothing than doctors looking for the slightest irregularity.

And we're all irregular in some way. What's important from a clinical perspective is how these incredibly small and sometimes large irregularities fit together.

The slightest feature can lead you into a completely new diagnostic direction. And, as you're about to see, the smallest detail can end up completely changing the way we view the world.

HE WAS beautiful in every way. And sleeping quietly in his Bugaboo stroller, Ethan looked pretty much like any other adorable baby.[7]

We all have a unique journey of development, but most of us share a common course of travel. This journey is paved and shaped by our environmental and genetic circumstances. And it always begins with the breathtaking beauty of an infant—small and vulnerable, yet full of so much potential.

The child sleeping before me had all of that. And although I didn't know it at the time, he was also unlike any baby I'd ever seen. Actually, he was unlike every other baby that's ever been born.

It's important to note that all of Ethan's fetal ultrasounds had been normal. Several months back, when his mother asked whether she would have a boy or a girl, her obstetrician had glided a wand through the blue ultrasound goo spread across her swollen belly and taken a peek between the unborn child's legs.

"It's a boy," she'd said.

And by all appearances, she had been right.

When he was born, Ethan did have one potentially concerning but not altogether uncommon trait. In most boys, the urethral opening—the place they pee from—is somewhere near the center of the head of their penis. But Ethan had hypospadias, which meant that the location of the urethral opening wasn't where it usually is but rather more toward the scrotum.

About one in every 135 boys is born with some form of hypospadias, from urethral openings down near the scrotum to all the way up near where it is on most boys—and it's generally quite fixable.[8] In most cases the correction is considered cosmetic, though surgeons sometimes must sacrifice the foreskin to make the repair. Sometimes parents decide that a slight, cosmetic case of hypospadias doesn't justify an operation. But in more severe cases, where a boy won't be able to stand to pee but will have to sit instead, surgery is often deemed to be important for psychosocial reasons.

As long as there's no blockage to the flow of urine, though, the surgical procedures to repair hypospadias are not done on an extremely urgent basis. So within minutes of his birth, when his condition was first noticed, Ethan's parents were informed and counseled about their options. And following all the usual first-day baby checks, they'd been sent home with the advice that they need not worry and could schedule a follow-up consultation with the surgical team to address his hypospadias in a few months.

Ethan's parents did worry, though, particularly as the months went by and their son clung to the bottom percentile for height and weight. They wanted to understand more about what they could do to get him up to size. But what started as a routine appointment to check his growth quickly turned into a puzzle of global proportions.

Given Ethan's size and seemingly benign physical trait, a common genetic test called a karyotype was ordered. In this test, a few of Ethan's cells were extracted, placed in a petri dish, encouraged to grow, and then treated with a special stain to help give contrast to his chromosomes.

That's when it began to become clear that Ethan was a little unlike other boys and men before him, who have all inherited a Y chromosome from their fathers. Though rare, it's not unheard of for a child who is genetically a girl to develop as a boy when a very small piece of the Y chromosome that contains a region called *SRY* (which stands for sex-determining region Y) is inherited. When this happens, a person's entire course of development can be shifted toward the male road instead of the female one.

In search of this little piece of *SRY*, the next step we employed in Ethan's case was called FISH (fluorescence in situ hybridization). The FISH test involves using a molecular probe that binds only to parts of the chromosome that are complementary.

What we expected to see in Ethan was that the FISH for the *SRY* region would be positive, as it is in other cases that present in this way. But it wasn't. In fact, it's not only that Ethan didn't inherit a Y chromosome from his father, he didn't even get a microscopic trace of one. And that didn't leave us with many known genetic explanations as to how Ethan could have turned into a boy.

Actually, according to the genetics textbook sitting on my desk, he really should have been a girl.

"IT'S A boy!" That's what Ethan's parents, John and Melissa, had longed to hear. And when they did, they were thrilled.

So was nearly everyone else in their extended family, especially John's parents, who were first-generation immigrants from mainland China. Even before the one-child policies took effect in their country, the birth of a boy was considered fortuitous—and so they were especially excited by the news that Melissa was pregnant with a boy.

And maybe just a little overprotective. At least once each day, Melissa would get a call at work from John's mother inquiring about her state of health and reviewing, as per their family's cultural traditions, what she should and shouldn't be doing, thinking, and eating. The long list of banned foods included two of Melissa's favorites: watermelons and mangoes.

That's not all. Melissa was also instructed to avoid ever leaving sharp objects, such as scissors or knives, on the bed—not only because she might accidentally cut herself, but also because John's mom had been raised to believe that such actions were bad luck and harbingers of undesirable omens that could cause the baby to have a "cut lip," what we call today a cleft lip or palate.

Melissa was not particularly superstitious, but in an effort to avoid any unnecessary familial conflicts she did her best to play along. There was one area, however, where she felt she needed to draw the line, at least secretly. As her pregnancy progressed, Melissa was insatiably craving watermelon. As long as she managed to keep the big green rinds and little black seeds hidden when her in-laws came over, she figured everything would be fine. When her mother-in-law happened to "volunteer" to take out the trash and found some rinds and that distinctive red juice at the bottom of the garbage bag, a tremendous fight ensued. Nothing Melissa could say would alleviate her mother-in-law's anger. Ultimately, she simply apologized and promised to stay away from all those "killer fruits" until long after her delivery, while quietly pledging to herself to be more careful about where she'd dump the evidence next time she had a secret snack.

Even though she knew her mother-in-law's fears were outlandish, when I shared with Melissa the news of her baby's genetic exceptionality, it made her wonder aloud if there could have been any truth to all those family superstitions. And while I'd never heard of anyone with this particular concern about watermelons before, her anxiety wasn't unusual at all.

The first question I hear from parents whose children have genetic conditions is often, "Doctor, is there anything I could have done to cause this?"

In situations like this I feel obliged to help alleviate the misplaced guilt that parents may feel. So instead of talking about all the possibilities at play for "what went wrong," I try very hard to frame the discussion in terms of what we know that has been scientifically established.

Of course, this usually requires that I have some idea. And in Ethan's case, at least at first, I had absolutely no clue.

ONE OF the possibilities that was brought up early on in Ethan's case was congenital adrenal hyperplasia, or CAH, a group of genetic conditions (caused by a handful of genes) that can make females look externally male. People with CAH don't naturally make enough of a steroid hormone called cortisol. When that deficiency is recognized by their bodies, their adrenal glands are stimulated to try to make more. The problem is, though, that's not all that gets made. More sex hormones can be produced as well.

In some cases of CAH, a version of the gene called *CYP21A* can cause girls and young women to develop bad acne, excessive body hair, and a large clitoris that can, in certain circumstances, look like a penis at birth. This is why CAH is one of the most common causes of ambiguous genitalia, making female babies appear more male.

The excess of androgens, caused by inheriting this gene, also interferes with the normal ovulatory cycle and prevents some of these women from being able to get pregnant. About one in 30 Ashkenazi Jews, about one in 50 women of Hispanic descent, and lower ratios of women from various other ethnicities have inherited genes that cause CAH, but many don't even know it.[9]

You don't need to undergo genetic testing to find out. There's a relatively simple blood test that can indicate whether a woman might be suffering from this form of CAH, but it's not always ordered. As a result, a lot of women spend years receiving ineffective fertility treatments, not to mention spending thousands of dollars, before learning that the condition preventing them from becoming pregnant is not actually a fertility problem at all but rather a genetic disorder that can be easily treated with a drug called dexamethasone.

But what about Ethan? Could his case be an unusually pronounced form of CAH? After a short discussion, we quickly crossed that possibility off the board. Genetic mutations that cause CAH can cause virilization in girls, even to the extent that they appear male at birth, but there's one thing they can't do, and that's make testes. As a visual inspection and testicular ultrasound confirmed, Ethan did indeed have two normally formed testes.

There are a few even rarer conditions that can cause XX-sex reversal of this type, but none of them lined up with what we were seeing in Ethan. Slowly but surely, one by one, we moved from likely to unlikely for every possible known cause for Ethan's condition and crossed them off the list.

Ultimately, our group coalesced around an idea made famous by Sir Arthur Conan Doyle's Sherlock Holmes: "When you have eliminated the impossible, whatever remains, however improbable, must be the truth." But as we whittled away at the impossible, what remained seemed so incredibly improbable that it took a long time for us to accept that it might indeed be true.

Maybe we've just been wrong about sex all along.

FOR A very long time the dogma has been that while chromosomally we may be male or female, developmentally we all start out the same. If we inherit a Y chromosome, or even a very small part of it, we take a detour toward maleness. In the absence of that, though, we'd all continue to head down the genetic path of being a female.

But in Ethan, as we saw, that wasn't the situation. So we began to suspect that the conventional genetic wisdom was indeed wrong.

Like one of the early spy satellites that first orbited the earth, most of the information gleaned from early genetic karyotype tests was grainy and lacked good resolution. It was essentially a mile-high glimpse of our packaged genome.

But even going back many decades, what the test could tell us was whether large sections of the arms that make up each chromosome were present.[10] In a way, performing a karyotype is like walking into an antiques store and staring at a bookshelf housing an encyclopedia. With a rather quick glance you can count the numbered volumes and see whether each one is present. The same goes for a karyotype, which provided a quick snapshot of whether each of our 46 chromosomes was present, though it would be impossible to say at that point whether all the pages upon which our genes are "printed" were inside, safe and intact.

In recent years, the resolution at which we can study genomes has increased phenomenally. We can now also use a detailed type of investigation, called a microarray-based comparative genomic hybridization in which we essentially "unzip" a person's DNA and then mix it together with a known sample of DNA. By comparing the two, we can identify small stretches of DNA that are either missing or duplicated. This accomplishes the same goal as a karyotype, but at an incredibly more detailed level.[11]

However, if you'd like to get even more information, down to the single letter spelling of your genome—to the point that we can see not just your chromosomes but look for rare changes in the sequence of each of the billions of individual nucleotides—adenosine, thymine, cytosine, guanine—then you need to sequence your DNA.

When it came to Ethan, we found one thing in particular that we didn't expect: He had a duplication of a gene called *SOX3* that's found on the X chromosome. Babies who develop into girls have two X chromosomes, so you'd expect them to have two copies of the *SOX3* gene. And they do, but usually one of their X chromosomes is randomly turned off, or "silenced," in every cell thanks to the product of a gene called *XIST*. Interestingly, Ethan's duplication would provide an extra opportunity for the gene *SOX3* to be expressed from nonsilenced X chromosomes. As we saw in an earlier chapter about gene dosage, where Meghan inherited extra copies of a gene that metabolized codeine, having extra numbers of genes can change or alter the overall amount of the protein product, which in Meghan's case caused a fatal overdose from codeine.

As it turns out, having an extra copy of the *SOX3* gene was significant for Ethan because it shares about 90 percent of its nucleotide sequence with the *SRY* region—a small piece of the Y chromosome that is a crucial signpost on the journey to becoming a male. The similarities are so significant that it's likely that *SOX3* is a genetic ancestor to *SRY*. The main difference: *SRY* exists only on the Y chromosome, while *SOX3* exists on the X chromosome.

As Sherlock might say: The game was afoot.

LIKE AN old baseball player coming out of retirement for one more game, it now seems clear, thanks to Ethan, that the *SOX3* gene has the ability to pinch-hit for *SRY*. And when put in the right place, at the right time, and in just the right circumstances, it can create a boy out of a girl, regardless of whether a Y chromosome is present or not.

Today, we know of a small number of other people with a similar, though not identical, genetic makeup as Ethan's. To further complicate things, what we've also learned is that some people who, like Ethan, have inherited a duplication of the *SOX3* gene and a "female" XX chromosomal complement have been found to develop as anatomically normal females.

So why is Ethan so different?

If you had told a geneticist 35 years ago that you could transform a slim brown mouse into a fat orange one and make that change heritable by giving it folic acid that turns its genes off and on, you likely would have been laughed at.

As we come to better understand the new and fast-changing genetic landscape around us, we are being forced to keep an open mind. Jirtle's agouti mice are just one very small example of the power of a singular environmental factor on the genome.

Our lives, of course, are seldom as singularly influenced as a lab mouse's life can be, a humbling reminder of the multitude of interactions across a very vast spectrum of variables that are occurring beyond our technological—and even intellectual—grasp.

The truth is, with all our advanced genetic tools, we still don't know exactly why Ethan turned into a boy, while others who've inherited a similar genetic makeup stayed the developmental course and became girls. But we know that in many other situations—Adam and Neil, the monozygotic twins with NF1, for instance—it doesn't take much to push our genetic expression or repression one way or another to forever change the course of our lives.

We've only just scratched the surface of the broad spectrum of genetic and epigenetic factors that influence our sexual development. And yet for most children like Ethan, the impact is still being felt in a very binary way. Boy or girl? He or she? Pink or blue?

But it doesn't have to be this way.

I FIRST encountered a *kathoey* while I was taking part in an HIV-prevention program with the Population and Community Development Association, or PDA, a nongovernmental organization operating in Thailand.

Her name was Tin-Tin, and she worked every night just a few steps away from where I was manning an educational booth in Patpong, Bangkok's world-famous red light district. One of PDA's goals in Thailand was to increase the use of condoms to help prevent the spread of HIV. This was, of course, especially important among the city's sex workers.

Tin-Tin's goal, on the other hand, was slightly different; to entice as many paying customers as possible into one of the local clubs that featured burlesque-style sex shows.

Even in heels, she was rather tall for a Thai woman, and in a place where sex workers congregated like bees in a hive, maybe because of her height, she stood out.

Patpong got its start on what was then the outskirts of Bangkok in the late 1940s but really hit its seedy stride during the Vietnam War, when hundreds of American GIs would spend their leave days and dollars doing the kinds of things that soldiers have always done. Today, though, the place has the feel of a tourist trap—a never-ending Mardi Gras–infused flea market and sexual playground.

Girls like Tin-Tin haunt the entrances of the clubs, either as employees who are trying to get foreign men and sexually adventurous couples to come in, or as self-employed entrepreneurs, seeking to entice those coming out to spend a little more money on a bit more fun.

For days, she eyeballed my educational booth, but she didn't come over to the table until one night when there was a sudden downpour and she skipped over—rather gracefully given the wet streets and her seven-inch heels—and ducked under a nearby awning.

She picked up one of the flyers prepared by the organization I was working for and casually flipped it over to the side that was printed in Thai.

"So, you married?" she asked in remarkably good English—and in a much deeper voice than I'd expected.

The storm lasted 30 minutes or so, and we spoke until it passed. That half-hour period with Tin-Tin was incredibly informative.

Here is some of what she revealed. There are about 200,000 people in Thailand who are considered *kathoey*—which many Thais, even socially conservative ones, consider to be a "third sex." Some are cross-dressers. Others are preoperative transgendered people. Others have surgically completed the full male-to-female transition.

And no, they're not all sex workers. *Kathoey* individuals work in every facet of Thai society, from garment factories to airlines, and even to the Muay Thai boxing ring. It's true: Arguably the most famous *kathoey* is a champion fighter named Parinya Charoenphol, a former Buddhist monk who pursued a Muay Thai career in order to raise enough money to pay for gender reassignment surgery. She would sometimes arrive in the ring wearing makeup and, after promptly dispatching her opponent, give him a postmatch kiss.

None of this is to say that *kathoeys* don't face a significant amount of discrimination in Thailand. They do. For one thing, there is no mechanism for changing one's legal gender from male to female, even for those who might indeed be genetically female. In a nation that conscripts about 100,000 young men into military service each year, this has caused some problems in the past.

Those seeking gender reassignment have other problems, too. The process in Thailand is relatively inexpensive by Western standards, which is why that nation is one of the most popular places in the world for people to go for sex change operations. But although it's cheaper, it's still out of reach for most Thais. Desperate, many *kathoeys* turn to prostitution to fulfill the dream of getting surgery.

And that was Tin-Tin's story. She was born into a poor farming family in the northeastern city of Khon Kaen and moved to Bangkok when she was 14 to earn a living. She was 24 when we met and still trying to put enough money away for an operation that, she had long since come to accept, might never actually happen. Faithfully, each month, she was also sending money home to her mother and father. "Where I'm from it is expected that sons take care of their parents," she told me. "Although I am more of a daughter to my mother and father now, I still feel that responsibility."

Over the next few weeks of occasional conversations with Tin-Tin, I learned a lot more, and I was accepted into what became her on-going dysmorphology course on the best ways to recognize a *kathoey*, which was fascinating.

"Take me," she said one night. "The best place to start is with the height. That's your first clue."

She was right. Across all ethnicities, genetically speaking, males tend to be significantly taller than females.

"Okay," I said, pointing to a shorter girl standing in front of a bar across the way. "What about that girl over there?"

"*Kathoey*," Tin-Tin said. "Look at her throat—you can see a big—what do you call this thing?" She tilted her head back and pointed to her throat.

"Adam's apple," I said.

"Yes—that," she said. "That's clue number two."

Again, she was genetically correct. The Adam's apple, known technically as a laryngeal prominence, is the result of male hormones that change the expression of genes during puberty, triggering tissue growth.

"Well, the first clue for me was your voice," I said.

"People can be so easily fooled by voices," she said, lifting her voice two octaves, overriding the deeper vocal tone from her own Adam's apple.

"Okay," I said, pointing to yet another girl, this one a regular visitor to my booth. "What about Nit? She's short. I've never noticed an Adam's apple on her. And she's got a high voice."

"*Kathoey*," Tin-Tin said.

"Are you sure?"

Tin-Tin looked at me and smiled knowingly, always the patient teacher.

"Of course, you can tell—look at her arms when she walks," she said. "See her arms? So straight, like a man. You're not looking at a real lady. She was born a boy. She's had surgery everywhere—lucky girl—but the elbows never lie."

What Tin-Tin was referring to was the carrying angle, the ever-so-slight way in which a female's forearms and hands turn away from the body when the arms are bent at the elbow. You can check it out for yourself if you stand in front of a mirror and simulate carrying a tray with your arms bent.

Don't be too concerned, though, if you find it more pronounced on yourself and you happen to be male. Tin-Tin's advice was sound—the bigger the carrying angle the more likely you're a female—but like many of our body parts, there's significant variability.

THAILAND ISN'T the only nation where a nuanced view of gender prevails.

Up until 2007, homosexual relationships were illegal in Nepal. But by 2011, the small south Asian nation of about 27 million residents made history as the first country in the world to conduct a census in which they counted not just men and women but a "third gender" including people who did not feel they fit neatly into either category.

In nearby India and Pakistan, a group known as *hijras*—physiological men who identify as women (and who sometimes submit to castration)—have also gained special recognition. As early as 2005, Indian passport authorities began to permit *hijras* to be uniquely identified as such on their documents, and starting in 2009, Pakistan followed suit.

Critical in all of these places is the idea that gender identity—or a lack thereof—is not a matter of choice. That doesn't impact in the slightest the prejudice that many people still unfortunately face, but it does set the stage for these relatively conservative societies to at least legally acknowledge and provide some measure of protection for those who don't fit into the classic binary roles of gender.

It's important to recognize that we're not talking about individuals and groups that have acquired a more liberal and modern idea of fluidity regarding gender from Western societies. *Hijras*, in particular, have a 4,000-year history in both India and Pakistan.[12]

Castration is also certainly not only a south Asian phenomenon. It spans dozens of cultures, including several relatively modern Western ones. In Italy, for instance, hundreds if not thousands of young boys were relieved of their testes between the sixteenth and nineteenth centuries for the cause of music. These boys became known as castrati.

Gizziello, Domenichino, and Carestini are far from household names today, but in the eighteenth century these castrati—who combined male lung capacity with female range, thanks to voices frozen in prepubescence—were Italy's A-listed singing stars. George Frideric Handel had a particular affinity for them; he wrote several operas, including Rinaldo, with castrati singers in mind.

Today there are only a few known recordings of a castrato, all of them made by Thomas Edison of the singer Alessandro Moreschi, who held the post of first soprano of the Vatican's Sistine Choir for three decades until his retirement in 1913.[13] Moreschi died in 1922 at the age of 63, an age that today would be quite young but, during that era, was more than a decade longer than the average life expectancy in Italy.

That might not be a coincidence. In addition to their distinctive voices, research into the lives of eunuchs who worked in the imperial court of Korea's Chosun Dynasty shows they lived decades longer than others who worked in the palace, including members of the royal family themselves, a phenomenon that researchers have suggested is evidence that male sex hormones such as testosterone might be damaging to cardiovascular health or weaken the immune system over time through modifications in both genetic expression and repression.[14]

I'm certainly not advocating castration as a way to tack a few extra years onto your life. What I am suggesting, however, is that our sexual biology isn't just about genetic sex, but rather the unique combination of genes, timing and the environment. As we keep seeing, people who fall away from the norm, for whatever reason, have a lot to teach the rest of us.

That's not just the case for the one-in-a-billion cases like Ethan, but also for hundreds of millions of people around the world who don't conform, genetically, biologically, sexually, or socially, to the rigid and traditional view of masculinity and femininity.

AS WE keep learning, our genes are incredibly sensitive. If it's a change in your diet, exposure to sunlight, or even bullying, our lives are continually informing our genetic inheritance. And when it comes to the timing of genetic expression or repression, it often doesn't take much to tip the scales.

In Ethan, after all, it didn't take an entire encyclopedia set, or even a single volume of genetic material, to turn him from a girl into a boy. All it took was a little extra genetic expression at just the right moment during his development. And so Ethan, with just his little extra dash of *SOX3*, forever and completely altered many of our perceptions about how we develop.

You've probably heard the saying "What lies behind us and what lies before us are tiny matters compared to what lies within us."[15] It's certainly a nice sentiment. But what we're now learning is that the tiny matter within us has a whole lot to do with what's behind us—and what's before us, too. In ways we previously could not have imagined.

Our cultural milieu can also have a significant impact on our sexual landscape. Consider again what happened in China, for instance, as ultrasounds provided a basic, binary snapshot of fetal development to more and more people, giving parents who would prefer boys a chance to eliminate girls by the millions. Remember, this was not what medical sonar was originally developed for. It was intended to help bring life into the world.

Today, the way some Chinese parents are using prenatal ultrasounds to choose boys over girls makes many people in the West feel uncomfortable. And yet we now live in a world in which gender is only one of many other things that can be chosen or eliminated before conception or during pregnancy using genetic testing.

Are we ready for a world in which children like Ethan, Tin-Tin, Richard, Grace, and all of the other people I've introduced you to in this book—not to mention the millions upon millions of others who exist outside of our social, cultural, sexual, aesthetic, and genetic norms—could be identified genetically and, like a submarine in the Caribbean, eliminated?

As we're about to see, by striving for even greater genetic perfection we might be eliminating a lot more than just millions of people who don't fit the societal norms we've created. We might actually be eradicating the very solutions to the medical problems we're working so hard to solve.

CHAPTER 11

Putting It All Together

What Rare Diseases Teach Us about Our Genetic Inheritance

By now you're very likely more attuned to all of the amazing, seemingly inconsequential genetic occurrences that have to happen—in just the right order, at just the right time—for a baby to be born.

And then, for that child to make it through her first day of life. And his first week. And their first year.

On and on it goes.

Through puberty. Into adulthood and parenthood. Through the changes of middle age. And as we learned in a previous chapter, against all the biological, chemical, and radiological influencers that conspire on a daily basis to change our genes.

However, it's the moment-to-moment biological events we may be missing. From the beating of your heart, to your lungs stretching out to fill with air at every breath, most of your biological life and its genetic consequences happen in the shadows. It's mostly in the extremes of physiological excess that you're reminded that your heart has likely never stopped beating since before you were born. When it's racing because you're aroused, nervous, or even exercising, your attention shifts to what is happening inside your body—but you may not often reflect on how a specific change is orchestrated by, and simultaneously impacting, a multitude of genetic and physiological mechanisms. As we've seen, our genomes exist in concert with the environment in which we live, responding moment by moment, by expression and repression, to what we need, when we need it.

Some of these events may be as mundane as the need for the creation of molecular machinery, in the form of an enzyme, that helps you digest your breakfast. Other moments may be more significant, requiring your genome to provide the template for proteins such as collagen, that are used for structural support or scaffolding, which helps you to heal and recover from the physical trauma of surgery.

It's unfortunate, I think, that whenever things are humming along seamlessly we spend most of our waking days blissfully ignorant of the details of the genetic underpinnings of our own inner workings, unaware that even at rest our bodies are in a constant state of motion. All too often, it's only once something has gone terribly wrong for ourselves, or someone we love, that we start to become somewhat more attuned to all the inexplicably complex and mind-bogglingly enigmatic things that had to happen, and need to keep happening, day after day, to get any of us from conception to birth to wherever we are at this very moment.

Like shadows moving behind a rice-paper screen, we do occasionally catch glimpses of our inner workings. We feel our pulse race when we're excited. We notice a cut scab over, then slowly disappear altogether. Through it all we are oblivious to the hundreds if not thousands of genes being continually expressed and repressed to make it all happen seamlessly until the inevitable happens.

As with a pipe that starts to leak in our home, we don't really give much thought to what's behind our walls or under our floors until they crack or burst. And then, when they do, it's just about *all* we can think of.

Life is like that. For the most part, our bodies don't ask for much in return for our continued existence. A few thousand calories a day, a little water, and some light exercise. That's it. The only payment required to maintain our precious lives.

Our bodies can even help us along, for the most part, like a fairly unobtrusive personal trainer or nutritionist. Molecular signals are ordered up that gently (or at times not so gently) remind us to eat, drink, and sleep. In releasing these little messengers, our bodies urge us to behave. But it's always a precarious sort of balance.

And if we ignore these demands, or if we do not have the means to satiate them, our bodies become restless until their needs are met (just think about the last time you needed, but couldn't find, a restroom). It all happens so effortlessly that most of us, for most of our lives, live in a state of almost complete physiological and genetic ignorance.

It's hard to recognize what's going right until something goes a little off. And then as we're about to see—almost as though you've removed a blindfold you had no idea that you were wearing—it all becomes crystal clear.

THERE'S NO one exactly like you on this entire planet.

But let me be clear: Even though you're genetically unique (unless you have a monozygotic twin and even then your epigenome is likely to be very different), there are a lot of people who might be really similar to you.

Sometimes, though, what makes us different are very small genetic changes—like Ethan's in the previous chapter—that can significantly impact and change our lives. And some of these changes are so unique that it's extremely difficult to find anyone else on the planet who shares them. If you're a geneticist, finding and studying what makes a person unique can change how you see the rest of humanity. And if geneticists are fortunate to make that kind of discovery, it may even lead to a new treatment for millions of other people around the world.

Such can be the gift of rarity. By understanding what makes genetic outliers different, we get a totally unique perspective on our own lives. New ways of seeing our genetic selves, informed through a glimpse offered by someone with a rare genetic disorder, clears the way for medical discoveries and treatments for the rest of us.

Which is why I'd like you to meet Nicholas.

By many accounts, Nicholas was a young teacher. Given that his very existence was remarkably unlikely—he's one of an exceedingly rare number of people in the world with a condition called hypotrichosis-lymphedema-telangiectasia syndrome, or HLTS—we knew that we had a lot to learn from him.

Now, you wouldn't need to be a trained dysmorphologist to know, at a single glance, that there was something different about Nicholas. What you might need someone like me to point out, though, is that there's a known genetic basis for that difference.

With brilliant blue eyes and a face that seemed to be fixed in a perpetual state of contemplation, this good-looking kid could also break out into a smile so big and contagious that you couldn't stop yourself from matching it. He was a young teenager, yet something about his disposition gave you the impression of a wisdom that was well beyond his years.

So striking and transfixing were these characteristics that you might at first barely notice the other traits for which his syndrome was named: hypotrichosis, a lack of hair; lymphedema, a continual cycle of swelling; and telangiectasia, webbed blood vessels at the surface of the skin.

The absence of much hair (Nicholas had just a few ginger wisps on the top of his head) and the spidery veins that were subtly apparent on his skin were both largely cosmetic issues. That doesn't mean those issues were unimportant, but neither of them were life threatening. The swelling, though, was another matter.

Under normal circumstances, our bodies do a remarkably good job of methodically moving around the various fluids that collect in our tissues as we go about our daily lives. Sometimes, in response to infection or injury, the fluid stays a bit longer in one area. Almost everyone experiences this at some point in their lives—if you've ever had a sprained ankle or wrist, you know how this goes. A bit of swelling is a very normal part of the healing process and usually does the body good. But in the case of people with HLTS, the swelling doesn't happen in response to injury but as a continual symptom of what appears to be a compromised lymphatic system, which isn't healthy at all.

Although HLTS is extremely rare—fewer than a dozen people worldwide are affected—all of these symptoms combined are quite common among people who have it. But Nicholas was also suffering from renal failure, which made him in dire need of a kidney transplant. As far as we knew, that wasn't "normal" for the other people who have been identified with HLTS. And that's what sent us on a trip around the world looking for an explanation.

Like many journeys, this one started with a map. Rather than highway numbers and street names, this map included a particular genetic address that was found, as far as we knew at the time, only in Nicholas' genome. By lining up all the letters of these DNA sequences against the known genomes of people who don't have HLTS and then observing where they diverge, we could see that HLTS is an apparent consequence of mutations or changes in a gene called *SOX18*.

Sometimes I like to get friendly with the genes I study, and to do this, occasionally I give them nicknames. This one I like to call the Johnny Damon gene, for the once shaggy-bearded Red Sox player who wore number 18 in Boston and also in New York, once he defected to the other side of that storied rivalry.

The Yankees recruited Damon because they had some expectations about what he could do for their team. He was, at that point, a career .290 hitter over 11 seasons in the league, a consistent threat to steal bases, and a rock-solid force in the outfield.

As is true for our genes, when you know what a player has done in the past it becomes a lot easier to predict how he will perform in the future. In four seasons with the Yanks, Damon continued to hit near .290—but in his final season in the Bronx he struck out nearly a hundred times (an unfortunate personal record), stole fewer bases than he had in any season of his career, and tied for the lead in American League left fielders in errors. When he entered into free agency at the end of the 2009 season, the Yanks declined to re-sign him.

Genes work like this, too. Once we know what a particular gene does under normal circumstances, it becomes easy to set a benchmark and see when it's not performing as expected, and vice versa. So, in the case of *SOX18*, people with HLTS help to highlight the important work the gene normally does in helping the body develop the right lymphatic mechanisms to pull back any excess fluid that leaks out into and between the crevices of our tissue.

That's incredibly useful information. But, of course, it still didn't help us to understand why Nicholas was suffering from kidney failure.

Could HLTS and his renal failure have been just a coincidence? Certainly. There are, after all, people all around the world who suffer from two or more similar medical problems that aren't at all connected genetically. Maybe Nicholas was simply unlucky in this same sort of way. That didn't sit well with me. I felt a persistent pull to continue exploring how his particular *SOX18* mutation and renal failure might be related, especially given the lack of any explanation. And so, with Nicholas as our guide, we embarked on another genetic adventure.

WHEN WE come across a patient in whom we've been able to identify a specific mutation, it's helpful—and can even be vital—to know whether it's original or inherited. Therefore, one of the first things we do is check the DNA of the patient's parents, to see which parent the mutation was inherited from. If the parents do not have the same mutation in their genes, it might be a new genetic change, one that we call de novo. We can't immediately assume we're looking at an original difference because we also have to account for a common human foible—infidelity.

And that, as you might imagine, can lead down a potentially prickly and perilous path of parental altercations, particularly if the genetic condition we've observed is one that others should be alerted to as a matter of life and death.

In Nicholas' case, we couldn't find the mutated gene in either of his parents' DNA, even after we confirmed paternity. So, according to what I just told you, that would mean we were looking at a new or de novo mutation.

Except for one tragic thing. The year after Nicholas was born, his mother, Jen, became pregnant with another boy. Seven months into the pregnancy, Jen became very sick. An investigation into her condition revealed that her baby was in crisis. Which prompted an emergency in utero surgery that failed to save the child. An evaluation of the lost child's DNA showed that he had the same *SOX18* variation as his brother. Nicholas was not alone.

Did both of the boys somehow develop exactly the same new mutation? That's incredibly unlikely. Rather, I suspect that one of

Nicholas' parents might be carrying a mutation in cells within their reproductive organs. When we see this type of inheritance pattern—parents without a mutation who have more than one child with the same genetic mutation—we call it gonadal mosaicism.

Now that it was established how Nicholas might have inherited his *SOX18* mutation, I was ready to dig deeper. And when I did, one thing continued to really stand out: The few other known individuals with his condition were homozygous for the *SOX18* mutation, meaning they carried two copies of the mutated gene. Nicholas, though, inherited only one copy of a misbehaving *SOX18* gene, not two, which meant that he was heterozygous for that mutation. Unlike Nicholas, these other "carrier" parents did not have HLTS. Even though they were all heterozygous and had only one mutation in their *SOX18* gene just like Nicholas had. Which means that if we understood the genetics correctly, Nicholas should not have had HLTS.

Many times in genetics, trying to answer one question leads to five new ones. What we'd hoped for Nicholas was that all these questions would bring us closer to the reason for his kidney failure. As I stepped back to reassess his case I began to wonder if Nicholas' striking kidney failure could be caused by another condition, one that was genetically similar but distinct from HLTS.

Theories are one thing. Attempting to prove or disprove them is of course a completely different story. To do that, we would need to find another genetic needle in a haystack made up of seven billion individuals. Practically speaking, the chances that we would find another person with the exact same genetic mutation and the exact same symptoms as Nicholas were next to none. With those kinds of odds I couldn't help but fail. Which meant it was definitely worth trying.

And so I did what any good geneticist who's looking for answers does: I went on tour. While on the road presenting Nicholas' case at as many medical conferences as possible, I kept on hoping that someone would show up who had seen a patient with symptoms similar to those experienced by Nicholas.

Looking back, I'm not sure exactly what I was naively thinking, given that the odds of this actually happening were staggeringly not in my favor. But knowing that it might just help Nicholas as well as provide an immense amount of valuable new medical knowledge, it was at least worth a shot.

As we've seen over and over, understanding rare cases like Nicholas' has the power to impact and change our lives as well. Thankfully, there's an entire world of genetic researchers and physicians who are dedicated to getting to the bottom of these very complicated medical mysteries. And unbeknownst to me at the time, on a completely different continent, there was a team of devoted physicians and researchers who happened to be asking the very same questions about a patient remarkably similar to Nicholas. Against incredible odds, this patient of theirs, Thomas also happened to have HLTS.

Just like Nicholas, and unlike the other few people with HLTS who had inherited two mutations, Thomas was found to be carrying only one copy of a *SOX18* mutation. Crucially, and to my complete and utter surprise, he also had suffered renal failure, leading to a kidney transplant.

Most importantly—here's the part we still can't fathom—Thomas not only shared the same clinical features as Nicholas, incredibly he shared the exact same mutation of one of his *SOX18* genes.

When I finally saw a photograph of Thomas, the experience was absolutely surreal. There, on my computer screen, staring back at me late one night when I was alone in my office, was a man who could have been—no, I might have sworn he was—the 38-year-old version of 14-year-old Nicholas.

Both of them had the same regal, nearly hairless heads, the same almond-shaped eyes, the same full, red, and deeply bowed lips, and, most of all, the same kind and wise look—as though they'd been carved from the same material substance.

Given the incredibly difficult journey they'd both been on, perhaps in a way they had.

For the moment, there's still no answer to the mystery of how these two individuals, separated by age and some 4,000 miles, came to exhibit such a strikingly similar genetic condition, physical appearance, and medical course, which included kidney failure, that apparently no one else on the planet has.

That similarity, added to all the others, left us with only one conclusion: We were looking at an entirely new condition.

Now, the benefits for the next person to come along with HLTRS (the extra R is for "renal") are pretty obvious. Nicholas has received his new kidney, an amazing gift from his father, Joe, and he's recovered quite well from the surgery. He's also been getting good marks on his report cards. No small feat for a young man who has missed so much school due to medical appointments and hospital visits. He's also recently been opening up socially in a way that he hadn't in the past. Notwithstanding the fact that he's an incredible kid with an impressively supportive and loving family, those very real quality-of-life improvements can also be attributed to the close medical supervision and the multidisciplinary and specialized expert care he's received since his condition has been more precisely identified. And what has worked for Nicholas and Thomas will be the first thing tried the next time around. Not to mention that the next patient will know much sooner that he or she is not alone in the world.

Of course, what we're talking about here could be a one-in-a-billion sort of situation—if that. The next time around could be a long, long time away.

So what does that have to do with the rest of us?

Well, quite a bit, actually.

TODAY, THERE are more than 6,000 known rare disorders. When they are all grouped together, we find that these conditions affect as many as 30 million Americans.[1] That's roughly one in 10 people living in the United States, or more than the entire population of Nepal.

A good way to visualize this is to picture a football stadium in which almost everyone is wearing a white shirt, save for every single person in every tenth row—those people are all wearing red. Look around the stadium. What do you see? A sea of red.

Now imagine that everyone wearing a red shirt is also holding an envelope. And imagine that in every envelope there is a piece of paper with a sentence on it. And imagine that all of those sentences, put together, tell a story about everyone else in the stadium.

That's how genetic research into rare diseases works. We've already talked about how just a small number of people who carry a mutation in the *SOX18* gene can help us better understand the way it works to help the body build its lymphatic system.

And here's where Nicholas and Thomas can help the rest of us: Many cancers hijack the lymphatic system for their own benefit and spread. Mapping out how *SOX18* is involved in this process will offer a new and much needed target for treating certain types of cancer. It's also certainly possible that Nicholas and Thomas might help us better understand the role of *SOX18* in supporting healthy kidneys.

Which is why, above all else, we are indebted to Nicholas, Thomas, and the multitude of others with genetic conditions who assist us in our work. They're far more likely, given the history of medical discoveries, to be providing potential health benefits to others than to be reaping benefits from it themselves.

This certainly isn't a new concept, and it far precedes our modern understanding of genetic medicine. Way back in 1882—two years before Gregor Mendel's death—a physician by the name of James Paget, now considered to be one of the founding fathers of medical pathology, noted in the British medical journal *The Lancet* that it would be shameful to set aside those who are impacted by rare diseases "with idle thoughts or idle words about 'curiosities' or 'chances'."

"Not one of them is without meaning," Paget continued. "Not one that might not become the beginning of excellent knowledge, if only we could answer the questions—why is it rare? Or being rare, why did it in this instance happen?"

What was Paget talking about? Well, just consider the story of one of the most successful drugs in medical history to see how clearly the rare can inform the common.

WE NEED fat. When we don't eat enough of it, life can become quite unpleasant—not just from a gastronomical perspective but from a physiological one as well. Ultralow-fat diets can lead to poor absorption of fat-soluble vitamins such as A, D, and E, and have even been linked in some people to depression and suicide.[2]

But, like many things in life, it's not very hard to get too much of a bad thing. And the trade-off for a high-fat diet, for many people, is too much low-density lipoprotein or LDL. Having too much LDL cholesterol in your blood can lead to atherosclerosis, a term that comes from the ancient Greek words *athero*, meaning "paste," and *skleros* meaning "hard." "Hard paste" is a really good way to describe the plaques that can build up along some of our arterial walls. As that happens, these vital passageways become narrowed and less flexible—a deadly combination predisposing often-unsuspecting victims to heart attacks and strokes.

And this, unfortunately, is not a rare condition. Cardiovascular disease, or CVD, affects about 80 million Americans and is the number one cause of death in the United States, claiming the lives of some half a million people a year.[3]

But we might not understand much at all about CVD if not for a very rare genetic condition called *familial hypercholesterolemia*, or FH.

In the late 1930s, a Norwegian physician named Carl Müller began looking into this disease, which is essentially an inherited form of very high cholesterol. What Müller learned was that people who are born with FH don't build up a high level of LDL—they start their lives with it.

Now, we all need some cholesterol to function—it's the starting material our bodies use to create many hormones and even vitamin D—but if we have too much of it floating through our bloodstream we run the risk of dying from complications related to heart disease. For people with FH, that fate can come really early in life because they can't easily shift the LDL out of their blood and into their liver, like most of us do. The result is extremely high levels of cholesterol that becomes trapped within the circulatory system.

Under normal circumstances, our bodies use *LDLR*, one of the genes implicated in FH, to make a receptor that the liver uses to mop up LDL. Normally, that helps keep this type of cholesterol from building up in your blood, oxidizing, and harming your heart. But if you carry a copy of the *LDLR* gene that has a mutation that leads to FH, then the normal movement of cholesterol doesn't function, and all that fat is left in your cardiovascular piping to potentially run amok.

It's not uncommon for men who carry two copies of these mutations to die from a heart attack in their 30s, or even earlier. This can happen even if they're running marathons and following the healthiest diet imaginable.

What Müller could not have imagined back then was that he was helping set the conceptual stage for the development of one of the biggest blockbuster drugs in pharmaceutical history.

We've long known that high LDL levels in most people can be addressed with diet and exercise. But since that's not enough for people with FH, those following in Müller's footsteps were seeking another way to knock down the high levels of LDL associated with this rare condition. What they came up with was a drug targeting an enzyme called HMG-CoA reductase. This enzyme is normally involved in helping our bodies make more cholesterol while we sleep at night. Blocking this enzyme with a corresponding drug, it was hoped, might result in lower levels of LDL in the blood. You may have even heard of this class of drugs or are taking one of them right now.

Atorvastatin,[*] more commonly known by the brand name Lipitor, is one of the most popular drugs of the group known as statins. It became a blockbuster drug and is currently prescribed to millions of people around the world. Unfortunately for some of the people who inherited mutations that led to FH and played such a key role in advancing our basic medical understanding, Lipitor isn't as effective. A few promising new orphan drugs are now being approved for use in people with FH. Yet for some of them, the only real way to bring their LDL levels under good control is a liver transplant.

For many millions of others, though, Lipitor has been a literal lifesaver, helping people with elevated cholesterol avoid an early demise from coronary artery disease, even if their health problems are not just solely related to genetics but rather to an indulgent lifestyle.

---

* Atorvastatin was not the first statin to be developed but is one of the most widely known.

When it comes to medicine, the people who need it most—and deserve it most—often don't get it first. And sometimes they don't get it at all.

But as we're about to see, that's not always the case.

SOMETIMES, THE distance between an initial genetic discovery and an important treatment innovation can take decades. That, as we discussed earlier, was the case in the quest to find a cure for PKU, starting with Asbjørn Følling's discoveries in the mid-1930s and culminating with Robert Guthrie's work in making tests for the disease accessible to nearly everyone.

Sometimes, though—increasingly and excitingly—things happen a lot more quickly. That's the story of argininosuccinic aciduria, or ASA, a metabolic disorder that affects the urea cycle in which the body struggles to get rid of normal amounts of ammonia.

Does that sound familiar? Yes, ASA is very similar to OTC, the condition shared by Cindy and Richard. Much as in the case of OTC, people with ASA have trouble converting ammonia through the cycle of steps it takes to end up with urea.

People with ASA also often suffer from cognitive delays. At first, it was assumed that these neurological effects were the result of the higher levels of ammonia in their systems, as in Richard's case. But doctors soon realized that, in people with ASA, the developmental issues remained at play and seemed to worsen over time, even when consistently lower levels of ammonia were maintained.

Recently, though, researchers at the Baylor College of Medicine began to home in on another symptom that some people with ASA suffer from: an unexplained increase in blood pressure. They knew that a simple molecule called nitric oxide was incredibly important for keeping blood pressure down. They also knew that the enzyme responsible for causing ASA is a prime route in the pathway for the production of nitric oxide in the body.

With that in mind, the Baylor team set aside some of the issues related to ammonia and focused directly on giving ASA patients drugs that act as nitric oxide donors. Lo and behold, the patients showed some promising improvements in memory and problem solving. And, as an added benefit, their blood pressure normalized, too.[4]

That's very far from a cure, but rather than decades, this vital link took just a few years to establish and is already being used by some doctors to try to treat some of the long-term symptoms of ASA. It's also helping inform the quest to address the involvement of nitric oxide depletion, which might be occurring in a variety of other much more common conditions such as Alzheimer's disease, another reminder of how the rare can help shed light on a condition that in one way or another affects us all.

Often, the ways in which people with rare diseases might be able to help the rest of us seem quite obvious. As we saw previously, by starting with people who have a rare genetic condition like FH that causes high cholesterol and heart attacks and eventually working toward a drug treatment like Lipitor, physicians can now help millions.

My own journey of pharmaceutical discovery and development was anything but straightforward. Sometimes the road from an obscure genetic condition to a new treatment is not linear. My ongoing interest in studying rare diseases eventually led to my discovery of a novel antibiotic that I named Siderocillin. What makes this antibiotic innovative is that it works like a smart bomb, specifically targeting "superbug" infections.

Back in the late 1990s, however, I wasn't interested in antibiotics at all. I was intensely studying a condition called hemochromatosis. This genetic disorder results in the body absorbing too much iron from the diet, which in some people can lead to liver cancer, heart failure, and an early death. What my research into hemochromatosis taught me, though, was that I could use some of the principles from this genetic disease to create a drug that targets killer microbes instead.

According to the Centers for Disease Control and Prevention, more than 20,000 people die every year in the United States alone due to infections by superbug microbes. What makes these organisms so deadly is that they are resistant to many, if not all, of the current antibiotics in our pharmaceutical arsenal. This is why my drug discovery has the potential to treat millions of people and save thousands of lives every year.

But at the time when I first proposed my invention, there was no scientifically established linear relationship between hemochromatosis and superbug infections. In fact, many other researchers I was working with couldn't understand why I seemed to be studying two separate problems simultaneously—resistant microbes and hemochromatosis. Now they do.

The knowledge I gained from studying rare genetic diseases has led to my being awarded twenty patents worldwide, with human clinical trials for Siderocillin slated to start in 2015. This is the clearest example I can think of from my own professional sphere of the power of applying the knowledge gained from rare genetic diseases that affect only a few of us to new treatment options for the rest of us.

Rare genetic conditions can help in other ways, too. As we're about to see, they can also stop us from harming our children—all for the sake of a few extra inches.

IMAGINE THE freedom of escaping your genetic inheritance. Envision the possibility of leaving behind any genes that put you at risk for a multitude of cancers. Okay, there's only one small catch. You would need to have Laron syndrome.

Untreated, most people with this condition are typically less than 4 feet 10 inches tall, have a prominent forehead, deep-set eyes, a depressed nasal bridge, a smallish chin, and truncal obesity. We know of around 300 people in the world who have this condition, and about a third of them live in a small number of remote villages in the Andean highlands of Ecuador's southern Loja Province.[5]

And they all appear to be virtually immune from cancer.

Why? Well, to understand Laron syndrome, it's helpful to know a bit about another genetic condition—one that exists on the opposite side of the spectrum, called Gorlin syndrome. People with this disorder are susceptible to a type of skin cancer called basal cell carcinoma.* While basal cell carcinoma is relatively common among adults who have spent a good portion of their lives in the sun, people with Gorlin syndrome can develop this type of skin cancer in their teens and without much sun exposure.

About one in 30,000 people are affected by Gorlin syndrome, though many are thought to go undiagnosed. Usually you don't know you have it until you or someone in your family gets diagnosed with cancer. There are, though, a few visual dysmorphologic clues that are occasionally present and that you could probably easily identify. These include macrocephaly (a large head), hypertelorism (wide eyes), and 2-3 toe syndactyly[6] (webbed second and third toes). Other common diagnostic features include small pits on the palms and uniquely shaped ribs that can be seen on a chest radiograph or X-ray.

So why are people with Gorlin syndrome so sensitive to getting malignancies, such as skin cancer, without exposure to sun? To an-

---

* With around two million new cases diagnosed every year, basal cell carcinoma is actually the most common type of skin cancer in the United States, although not the most deadly. Of course, not everyone with basal cell carcinoma has Gorlin syndrome.

swer that question I need to tell you about a gene called *PTCH1*. Our bodies typically use this gene to makes a protein called Patched-1, which plays a crucial role at keeping cellular growth in check. But when a protein called Sonic Hedgehog* comes along in Gorlin patients, whose Patched-1 is not working properly, it releases the hold on growth that would usually be there and that makes cells free to divide. And divide. And divide.[7]

This, of course, is a problem, because as we've now seen many times, unrestricted growth is like cellular anarchy. And unfortunately, cancer can result.

Okay, so what does Gorlin syndrome teach us about Laron syndrome? Essentially, Gorlin syndrome represents, in a way, the genetic inverse to Laron syndrome. Whereas in one there is a promotion of cellular growth, the other experiences cellular growth restriction. Laron syndrome is caused by mutations in the receptor for growth hormone. This makes people with Laron syndrome insensitive or immune to it—one of the reasons they are often quite short.

Rather than the cellular anarchy seen in people with Gorlin syndrome, in those with Laron syndrome there is a stranglehold on growth, a form of extreme cellular totalitarianism.

Now, politically you might have some reservations about totalitarianism as an ideology, but from a purely biological perspective it has been incredibly successful. If it weren't, you wouldn't be here reading this right now. Neither would I. Neither would any of the other multicellular organisms on this planet.

Because, like you and me and all the other multicellular creatures, we are the product of biological totalitarianism that promotes cellular obedience at all costs, an obedience enforced by receptors on the surface of any potentially misbehaving cells that result in cellular

---

* In case you're wondering, Sonic Hedgehog is in fact named after the Sega video game character.

seppuku or *hara-kiri*—a programmed type of cell suicide known as apoptosis.

Like samurai warriors who become dishonored, cells that have the impudence to have greater aspirations than just being one in a crowd of many trillion are programmed and occasionally commanded to end their own lives. By this same mechanism, cells that are infected with pathogens can also sacrifice themselves to protect the body from microbial invaders. It's also the same mechanism we learned about previously that frees our fingers and toes from the webbing that's there during development. If those cells don't die—as happens in some genetic conditions—you can end up having mittens for hands.

Which is why, as in all things, equilibrium is crucial. Processes that restrict growth need to be constantly balanced with times when growth is needed. Just think about every time you've sustained an injury, be it a simple cut or a much more serious accident. Consider the entire repair and remodeling your body did—automatically. All of this is a process of the balance being struck, millions upon millions of times each day, between cellular life and death.

Would you want to mess with that balance?

Well, you or someone you know probably already has.

BEING TALL has its benefits. Taller kids are bullied less and get more playing time on the sports field. Taller adults, research has shown, are thought to more easily ascend to jobs of higher status and authority and earn more, on average, than their shorter coworkers.[8]

Of course, there are exceptions. Among the most famous is Napoléon Bonaparte. As it turns out, the world's most famous vertically challenged person might not have been that short after all. Back around the turn of the nineteenth century, French inches were a little longer than British ones. So while the Brits, who weren't exactly Napoleon's biggest fans, put his height at five feet and not much more, he was likely closer to five feet five, and may have been as tall as five feet seven, which was by no means short in his time.[9]

But whether they're French or English inches, when it comes to height, every one of them counts. And, let's face it, people who are able to reach the top shelf without a stepping stool can be just plain useful at times.

All of this is why short stature—or the perception of it—is the second most common referral to pediatric endocrinologists. It's not that parents wouldn't love their kids just as much if they maxed out on the shorter side of things, it's that height in our generation has become a real commodity. And after more than a half century of recombinant growth hormone (GH) therapy being made available for the small numbers of children with significant growth deficiencies, parents have now become quite aware that they can indeed have an impact on their child's height—and theoretically give them a leg up on their future.[10]

Today, there is an ever-growing list of conditions, some of which you've already read about in this book, for which GH, the manufactured version of human growth hormone, is prescribed. From Prader–Willi syndrome (the first human disorder linked to epigenetics) to Noonan's syndrome (the disorder I identified in my wife's friend Susan over dinner a few years ago), researchers are finding that more and more people may be able to benefit from an extra injection of GH here and there.

Some of these conditions are very serious disorders for which GH is an essential component in addressing the needs of sick kids. But in many cases, the administration of GH (typically through regularly scheduled injections) is specifically used to address matters of height alone. Idiopathic short stature, for instance, is a condition in which a child's height is more than two standard deviations below average, but without indications of any genetic, physiological, or nutritional abnormalities that we can identify. In other words, they're likely normal kids who happen to be really short.

And that's what troubles Arlan Rosenbloom. When I asked the University of Florida endocrinologist (who was one of the individuals instrumental in the discovery that Laron syndrome patients rarely if ever got cancer) if he had any concerns about giving children growth hormone, he answered with a single word: endocosmetology. That's what Rosenbloom (and a rapidly increasing number of his colleagues) somewhat derisively call the use of growth hormones for cosmetic purposes, including the desire to increase a child's height.[11]

But if GH has cleared all the regulatory hurdles (and there are many) for use in children, and epidemiological studies have not demonstrated an increased risk of cancer in children treated with GH, why should we be concerned?

Well, to answer this it might be helpful to look at something called insulin growth factor 1, or IGF-1, which is released after the body senses a surge of growth hormone. IGF-1 doesn't only promote vertical growth; it also promotes the survival of cells—and if you're trying to push a few more inches onto a child's short frame, that might be a good thing.

But before you allow your child to be treated with GH, consider this: IGF-1 is also thought to inhibit apoptosis—cellular suicide— and in the case of a group of cells gone rogue, that could be dangerous.

Or even deadly.

In Rosenbloom's view, giving children growth hormone just because they're a little shorter than other kids exposes them to an unnecessary risk—possibly to cancer down the road—one we may not be able to fully comprehend today but only in the decades to come. And he believes that decisions being made to treat children with GH are increasingly the result of market-driven campaigns by drug companies rather than decisions made for the health and long-term well-being of our children.

Today, the market for GH is worth billions, and millions are spent every year on marketing to advise worried parents that their precious child who may be on the short side requires a costly intervention for what might not be a real problem.

If people with Laron syndrome do not get cancer because their bodies can't respond to growth hormone, should we accept the risks and keep injecting our children with a synthetic version of the same hormone? If more parents learned about Laron syndrome, there's a good chance that, given the potential cancerous implications of growth hormone administration, they might be a little less inclined to use it.

WHEN LARON syndrome was first described back in the mid-1960s, there was no way to predict that so many years later it would be offering us a rare glimpse of immunity to cancer—or that studying any rare disease would lead to anything more than esoteric medical knowledge.

But as we've seen on this genetic odyssey, it's often the rare family who have genes that predispose them to high cholesterol (for example) who ultimately turn around and help us make medical breakthroughs for countless others. After all, studying families with hemochromatosis led to my discovery of a new antibiotic. We owe an immeasurable amount of gratitude to every person with a rare disease and to their families for these medical gifts.

Over the years I've met an incredible group of people with rare disorders. Still, I'd never presume to know what it's like to walk in any of their shoes—the truth is that no one can.

But my role gives me a unique perspective—indeed, a very close vantage point into the worlds of some of the toughest people I've ever met: Patients, parents, spouses, and siblings who have shown unbelievable courage in the face of a challenging diagnosis that tests their patience, compassion, physical endurance, and emotional fortitude.

Take Nicholas' mother, for instance. Over the years, Jen has developed a reputation as a "Kung-Fu Momma" for her resolute and steadfast advocacy on behalf of her son.

I mentioned this nickname to Jen once, and she brimmed with pride (and Nicholas laughed hysterically). And that's good—because the truth is that, as physicians, we really rely on parents like her to push us to go deeper and to think creatively about their children's conditions.

And then there's always the lesson and reminder about what it means to be thankful for all those seemingly inconsequential things that need to happen, day in and day out, that have brought you to where you are today. Things you don't even notice until the ever-so-rare occasion when something goes wrong. I'm not just talking about what's going on inside all our genomes but about what it means to be human. About what it means to live. To overcome. To love.

And that's not all. As we've seen now many times over, these amazing patients and their inspiring families can also help us diagnose, treat, and cure countless of other conditions. Being around them reminds me that I often stand to learn more from my patients than they can from me.

We all do.

Because hiding deep inside of everyone with a rare genetic condition is a secret that, if they choose to share it, might one day serve to cure and help every last one of us.

EPILOGUE

One Last Thing

We've covered a lot of ground, from the bottom of the Caribbean to the top of Mount Fuji, meeting genetically doped athletes, remarkable human pincushions, ancient bones, and hacked genomes.

We've also seen how our genes don't easily forget the trauma of being bullied, how a simple dietary change can turn workers into queen bees, and if you're not careful on that next vacation, how even a small indiscretion can effortlessly alter your DNA.

Through it all, we saw how our genetic inheritance can change and be changed by what we experience. We know that in our lives—as it is for all life on this planet—flexibility is key. And rigidity, as we've learned, can be the surprising enemy of strength.

Even a tiny change in the expression of your genome during development can reverse a person's sex. Ethan turned into a boy instead of becoming a girl, not simply because of what he inherited but because of a small change in the exquisite timing of his genetic expression. Remember, many others with genetic sequences similar to Ethan's develop into girls.

We've also explored how the understanding of the inner workings of your own DNA is a gift from people with rare genetic conditions—and we owe them much. Surprisingly, it is by understanding the limitations we've inherited that we're offered the best chance to transcend them. Knowing what to do with your genetic inheritance gives you the power to shape it.

This is why you may someday be talking to a friend who will tell you that she's been having more fruit and vegetables lately and it's been making her feel very bloated and tired. And you're going to remember Jeff the Chef. Maybe you won't remember what his condition is called (hereditary fructose intolerance), but you'll almost certainly remember something so much more important—that there's no universally perfect diet. As we've learned from Jeff, diets that are good for many of us can be deadly for some.

And maybe because of this book, after your children are born and one of them is a bit smaller than the others, you'll overhear someone talking about growth hormone therapy. You'll remember the genetic condition (Laron syndrome) that particularly affects a hundred or so people who live in the mountains of Ecuador. You just might recall that these people don't appear to get cancer because they are immune to growth hormone, and in that way you'll have information at your disposal that will help you make an informed decision.

Remembering Meghan, for whom a few extra copies of a single gene, *CYP2D6*, turned a prescription for codeine into a death sentence will give you the courage to speak up, not only for your child but for all those with rare diseases whose lives so vitally inform our collective medical knowledge.

That's what Liz and David are doing for little Grace. Her bones won't likely be as strong as most people's, but she's demonstrating every day both to me and to all those around her that her genome is not a completed book already written, edited, and published. It's a story that she's still telling.

Remember what that orphanage worker told them? "You are her destiny," she said. Not her genes. Not her brittle bones. The woman and man who decided they needed to be her parents and who gave her the gift of a completely new birthright. A new chance to survive, despite her genetic inheritance, and the opportunity to thrive.

As we're discovering, our genetic strength isn't just a matter of receiving the genes handed down to us from previous generations. It's derived from the opportunity to transform what we get and what we give.

And in doing so completely change the course of our lives.

NOTES

CHAPTER 1: HOW GENETICISTS THINK

1 Some names in this book have been changed and some identities, descriptions and scenarios have been altered or combined to protect the confidentiality of patients, friends, acquaintances, and colleagues, or to give clarity to an existing idea or diagnosis.

2 Although the price has come down significantly for both exome and whole genome sequencing, the time and cost associated with the interpretation of the data still needs to be considered.

3 There are some fundamental psychological principles at play here. For further reference, read J. Nevid (2009). *Psychology Concepts and Applications.* Boston: Houghton Mifflin.

4 M. Rosenfield (1979, Jan. 15). Model expert offers "something special." *The Pittsburgh Press.*

5 P. Pasols (2012). *Louis Vuitton: The Birth of Modern Luxury.* New York: Abrams.

6 The National Center for Biotechnology Information is a comprehensive and reliable public resource for information on all kinds of diseases, including Fanconi anemia: www.ncbi.nlm.nih.gov.

7 Rearrangements of the *PAX3* gene are also thought to be involved in some forms of rare cancers called alveolar rhabdomyosarcoma. S. Medic and M. Ziman (2010). *PAX3* expression in normal skin melanocytes and melanocytic lesions (naevi and melanomas). *PLOS One, 5*: e9977.

8 About one child in every 700 live births has Down syndrome.

9 Although not routinely employed today, analysis of fetal meconium can be used to test for gestational alcohol exposure by the presence of chemicals called fatty acid ethyl esters or FAEEs.

10 If having a fat thumb is something that should be kept hidden, what does that say to those of us who have even more severe and debilitating physical anomalies? This, to me, is an extremely sad statement about the lengths that marketers have gone to establish the idea of perfect people—and especially perfect women. See I. Lapowsky (2010, Feb. 8). Megan Fox uses a thumb double for her sexy bubble bath commercial. *New York Daily News*.

11 K. Bosse et al. (2000). Localization of a gene for syndactyly type 1 to chromosome 2q34-q36. *American Journal of Human Genetics, 67*: 492–497.

12 Marriage between relatives can increase the likelihood of genetic disorders anywhere from double the risk to higher, depending upon the ethnicity of the family.

13 Dysmorphology is a subspecialty of medicine that uses our anatomical features to understand our genetic and environmental history. If the terminology used by dysmorphologists excites you, I suggest reading Special Issue: Elements of Morphology: Standard Terminology. (2009). *American Journal of Medical Genetics Part A, 149*: 1–127. If you'd like to learn more about this fascinating field, start with *The Journal Clinical Dysmorphology*, a peer-reviewed collection of articles on cases and research related to this field.

CHAPTER 2: WHEN GENES MISBEHAVE

1 S. Manzoor (2012, Nov. 2). Come inside: The world's biggest sperm bank. *The Guardian*.

2 C. Hsu (2012, Sept. 25). Denmark tightens sperm donation law after "Donor 7042" passes rare genetic disease to 5 babies. *Medical Daily*.

3 R. Henig (2000). *The Monk in the Garden: The Lost and Found Genius of Gregor Mendel, the Father of Genetics*. New York: Houghton Mifflin.

4 In Mendel's original publication he used the German word *vererbung*, which we would translate in English as "inheritance." The use of the term predates Mendel's paper.

5 D. Lowe (2011, Jan. 24). These identical twins both have the same genetic defect. It affects Neil on the inside and Adam on the outside. U.K.: *The Sun*.

6 M. Marchione (2007, Apr. 5). Disease underlies Hatfield-McCoy feud. The Associated Press.

7 If you'd like to learn more about von Hippel-Lindau disease and support organizations, please see the following NORD website: www.rarediseases.org/rare-disease-information/rare-diseases/byID/181/viewFullReport.

8 L. Davies (2008, Sept. 18). Unknown Mozart score discovered in French library. *The Guardian*.

9 M. Doucleff (2012, Feb. 11). Anatomy of a tear-jerker: Why does Adele's "Someone Like You" make everyone cry? Science has found the formula. *The Wall Street Journal*.

10 You can listen to Leisinger play Mozart's piano at www.themozartfestival.org.

11 G. Yaxley et al. (2012). *Diamonds in Antarctica? Discovery of Antarctic Kimberlites Extends Vast Gondwanan Cretaceou Kimberlite Province*. Research School of Earth Sciences, Australian National University.

12 E. Goldschein (2011, Dec. 19). The incredible story of how De Beers created and lost the most powerful monopoly ever. *Business Insider*.

13 E. J. Epstein (1982, Feb. 1). Have you ever tried to sell a diamond? *The Atlantic*.

14 H. Ford and S. Crowther (1922). *My Life and Work*. Garden City, NY: Garden City Publishing.

15 D. Magee (2007). *How Toyota Became #1: Leadership Lessons from the World's Greatest Car Company*. New York: Penguin Group.

16 A. Johnson (2011, Apr. 16). One giant step for better heart research? *The Wall Street Journal*.

17 There are many papers published on this topic. Here is one I particularly enjoy reading: H. Katsume et al. (1992). Disuse atrophy of the left ventricle in chronically bedridden elderly people. *Japanese Circulation Journal, 53*: 201–206.

18 J. M. Bostrack and W. Millington (1962). On the determination of leaf form in an aquatic heterophyllous species of *Ranunculus*. *Bulletin of the Torrey Botanical Club, 89*: 1–20.

CHAPTER 3: CHANGING OUR GENES

1 This paper is cited by nearly a hundred others and stands out as a landmark: M. Kamakura (2011). Royalactin induces queen differentiation in honeybees. *Nature, 473*: 478. If you find bees as fascinating as I do, you might like to read this paper as well: A. Chittka and L. Chittka (2010). Epigenetics of royalty. *PLOS Biology, 8*: e1000532.

2 F. Lyko et al. (2010). The honeybee epigenomes: Differential methylation of brain DNA in queens and workers. *PLOS Biology, 8*: e1000506.

3 R. Kucharski et al. (2008). Nutritional control of reproductive status in honeybees via DNA methylation. *Science, 319*: 1827–1830.

4 B. Herb et al. (2012). Reversible switching between epigenetic states in honeybee behavioral subcastes. *Nature Neuroscience, 15*: 1371–1373.

5 Humans have two different versions, *DNMT3A* and *DNMT3B,* which have shared homology and similarity in the catalytic domain to the Dnmt3 gene found in *Apis mellifera*, the honeybee. If you'd like to read more about this, see the following paper: Y. Wang et al. (2006). Functional CpG methylation system in a social insect. *Science, 27*: 645–647.

6 M. Parasramka et al. (2012). MicroRNA profiling of carcinogen-induced rat colon tumors and the influence of dietary spinach. *Molecular Nutrition & Food Research, 56*: 1259–1269.

7 A. Moleres et al. (2013). Differential DNA methylation patterns between high and low responders to a weight loss intervention in overweight or obese adolescents: The EVASYON study. *FASEB Journal, 27*: 2504–2512.

8 T. Franklin et al. (2010). Epigenetic transmission of the impact of early stress across generations. *Biological Psychiatry, 68*: 408–415.

9 R. Yehuda et al. (2009). Gene expression patterns associated with posttraumatic stress disorder following exposure to the World Trade Center attacks. *Biological Psychiatry, 66*: 708–711; R. Yehuda et al. (2005). Transgenerational effects of posttraumatic stress disorder in babies of mothers exposed to the World Trade Center attacks during pregnancy. *Journal of Clinical Endocrinology & Metabolism, 90*: 4115–4118.

10 S. Sookoian et al. (2013). Fetal metabolic programming and epigenetic modifications: A systems biology approach. *Pediatric Research, 73*: 531–542.

CHAPTER 4: USE IT OR LOSE IT

1 E. Quijano (2013, Mar. 4). "Kid President": A boy easily broken teaching how to be strong. CBSNews.com.

2 Thankfully, these sorts of stories are quite rare. Nonetheless, this is an incredibly tragic tale. H. Weathers (2011, Aug. 19). They branded us abusers, stole our children and killed our marriage: Parents of boy with brittle bones attack social workers who claimed they beat him. *The Daily Mail.*

3 U.S. Department of Health & Human Services (2011). *Child Maltreatment.*

4 FOP was detailed in medical literature as far back as 250 years ago, but the disease's cause was a medical mystery until fairly recently. If you'd like to read more on FOP see the following paper: F. Kaplan et al. (2008). Fibrodysplasia ossificans progressiva. *Best Practice & Research: Clinical Rheumatology. 22*: 191–205.

5 Ali's family has raised an "army" for their daughter and others who suffer from FOP: N. Golgowski (2012, June 1). The girl who is turning into stone: Five year old with rare condition faces race against time for cure. *The Daily Mail.*

6 Today, keeping an eye on the big toes of people suspected of FOP is part of the standard dysmorphology exam: M. Kartal-Kaess et al. (2010). Fibrodysplasia ossificans progressiva (FOP): Watch the great toes. *European Journal of Pediatrics, 169*: 1417–1421.

7 A. Stirland (1993). Asymmetry and activity related change in the male humerus. *International Journal of Osteoarcheology, 3*: 105–113.

8 The *Mary Rose* remained on the seabed until it was raised in 1982. Ever since then, scientists have been racing to uncover the identities and life stories of the sailors aboard: A. Hough (2012, Nov. 18). *Mary Rose*: Scientists identify shipwreck's elite archers by RSI. *The Telegraph.*

9 If you happen to be interested in the heritability of bunions, please see M. T. Hannan et al. (2013). Hallux valgus and lesser toe deformities are highly heritable in adult men and women: The Framingham foot study. *Arthritis Care Research* (Hoboken). [Epub ahead of print.]

10 In any other context, a loaded backpack might be considered a torture device. See D. H. Chow et al. (2010). Short-term effects of backpack load placement on spine deformation and repositioning error in schoolchildren. *Ergonomics, 53*: 56–64.

11 A. A. Kane et al. (1996). Observations on a recent increase in plagiocephaly without synostosis. *Pediatrics, 97*: 877–885; W. S. Biggs (2004). The "epidemic" of deformational plagiocephaly and the American Academy of Pediatrics' response. *JPO: Journal of Prosthetics and Orthotics, 16*: S5–S8.

12 Before you invest in a cranial remodeling helmet please consider J. F. Wilbrand et al. (2013). A prospective randomized trial on preventative methods for positional head deformity: Physiotherapy versus a positioning pillow. *The Journal of Pediatrics, 162*: 1216–1221.

13 It's a fantastically fascinating fish. For more information please see J. G. Lundberg and B. Chernoff (1992). A Miocene fossil of the Amazonian fish *Ara-*

paima (Teleostei Arapaimidae) from the Magdalena River region of Colombia—Biogeographic and evolutionary implications. *Biotropica, 24*: 2–14.

14 M. A. Meyers et al. (2012). Battle in the Amazon: Arapaima versus piranha. *Advanced Engineering Materials. 14*: 279–288.

15 A very small genetic change that led to a lethal type of OI was just one of the first of many high-profile revelations of the power of a change in a single nucleotide. See D. H. Cohn et al. (1986). Lethal osteogenesis imperfecta resulting from a single nucleotide change in one human pro alpha 1(I) collagen allele. *Proceedings of the National Academy of Science, 83*: 6045–6047.

16 D. R. Taaffe et al. (1995). Differential effects of swimming versus weight-bearing activity on bone mineral status of eumenorrheic athletes. *Journal of Bone and Mineral Research, 10*: 586–593.

17 The photos and videos that accompanied this story about the space capsule landing show the three spacemen struggling with their sudden reimmersion in Earth's gravity. See P. Leonard (2012, July 2). "It's a bullseye": Russian Soyuz capsule lands back on Earth after 193-day space mission. *Associated Press*.

18 A. Leblanc et al. (2013). Bisphosphonates as a supplement to exercise to protect bone during long-duration spaceflight. *Osteoporosis International, 24*: 2105–2114.

CHAPTER 5: FEED YOUR GENES

1 F. Rohrer (2007, Aug. 7). "China drinks its milk." *BBC News Magazine*.

2 Which makes sense, given the fact that many people don't know how to cook much at all, let alone cook food that is tasty and nutritious. See this paper for more information: P. J. Curtis et al. (2012). Effects on nutrient intake of a family-based intervention to promote increased consumption of low-fat starchy foods through education, cooking skills and personalized goal. *British Journal of Nutrition, 107*: 1833–1844.

3 D. Martin (2011, Aug. 18). From omnivore to vegan: The dietary education of Bill Clinton. *CNN.com*.

4 S. Bown (2003). *Scurvy: How a Surgeon, a Mariner and a Gentleman Solved the Greatest Medical Mystery of the Age of Sail*. West Sussex: Summersdale Publishing Ltd.

5 L. E. Cahill and A. El-Sohemy (2009). Vitamin C transporter gene polymorphisms, dietary vitamin C and serum ascorbic acid. *Journal of Nutrigenetics and Nutrigenomics, 2*: 292–301.

6 H. C. Erichsen et al. (2006). Genetic variation in the sodium-dependent vitamin C transporters, *SLC23A1*, and *SLC23A2* and risk for preterm delivery. *American Journal of Epidemiology, 163*: 245–254.

7 If you'd like to read more, here's a paper that explores some of these ideas: E. L. Stuart et al. (2004). Reduced collagen and ascorbic acid concentrations and increased proteolytic susceptibility with prelabor fetal membrane rupture in women. *Biology of Reproduction. 72*: 230–235.

8 Jeff the Chef, whom we met in the introduction, found himself in this position when he followed his doctor's nutritional advice.

9 If you'd like to read more about the pharmacogenetics of caffeine intake, see: Palatini et al. (2009). *CYP1A2* genotype modifies the association between coffee intake and the risk of hypertension. *Journal of Hypertension, 27*:1594–601 and M. C. Cornelis et al. (2006). Coffee, *CYP1A2* genotype, and risk of myocardial infarction. *The Journal of the American Medical Association, 295*:1135–1141.

10 I. Sekirov et al. (2010). Gut microbiota in health and disease. *Physiological Reviews, 90*: 859–904.

11 Often there needs to be a waiting period of a few weeks for room to develop in the growing body cavity. A specialized temporary packing called a silo is constructed around the intestines to protect the baby's intestines during the wait. Although the silo can be very visually disconcerting to the parents and family of a child with gastroschisis, this period of time is necessary for enough room to develop to accept the intestines, so that they can be safely packed back into the body and the wall surgically closed and corrected.

12 N. Fei and L. Zhao (2013). An opportunistic pathogen isolated from the gut of an obese human causes obesity in germfree mice. *The ISME Journal, 7*: 880–884.

13 If you're interested to read more about this topic, see the following paper: R. A. Koeth et al. (2013). Intestinal microbiota metabolism of l-carnitine, a nutrient in red meat, promotes atherosclerosis. *Nature Medicine, 19*: 576–585.

14 S. A. Centerwall and W. R. Centerwall (2000). The discovery of phenylketonuria: The story of a young couple, two retarded children, and a scientist. *Pediatrics, 105*: 89–103.

15 P. Buck (1950). *The Child Who Never Grew.* New York: John Day.

CHAPTER 6: GENETIC DOSING

1 If you'd like to read more about cases like Meghan's, here's a good place to start: L. E. Kelly et al. (2012). More codeine fatalities after tonsillectomy in North American children. *Pediatrics, 129*: e1343–1347.

2 What happened in those intervening years? A lot of slow movement toward a lifesaving conclusion. Many times, unfortunately, that's how medical science works. See B. M. Kuehn (2013). FDA: No codeine after tonsillectomy for children. *Journal of the American Medical Association, 309*: 1100.

3 A. Gaedigk et al. (2010). *CYP2D7-2D6* hybrid tandems: Identification of novel *CYP2D6* duplication arrangements and implications for phenotype prediction. *Pharmacogenomics, 11*: 43–53; D. G. Williams et al. (2002). Pharmacogenetics of codeine metabolism in an urban population of children and its implications for analgesic reliability. *British Journal of Anesthesia, 89*: 839–845; E. Aklillu et al. (1996). Frequent distribution of ultrarapid metabolizers of debrisoquine in an Ethiopian population carrying duplicated and multiduplicated functional *CYP2D6* alleles. *Journal of Pharmacology and Experimental Therapeutics. 278*: 441–446.

4 Rose, who died in 1993, is a hero to many doctors and researchers—and rightfully so: B. Miall (1993, Nov. 16). Obituary: Professor Geoffrey Rose. *The Independent.*

5 Much as we know that the effects of codeine vary widely depending on a person's genetic inheritance, so, too, have we learned that the effects of just about every medical intervention can be very different from person to person, sometimes for the better and sometimes for the worse: G. Rose (1985). Sick individuals and sick populations. *International Journal of Epidemiology, 14*: 32–38.

6 See A. M. Minihane et al. (2000). *APOE* polymorphism and fish oil supplementation in subjects with an atherogenic lipoprotein phenotype. *Arteriosclerosis, Thrombosis, and Vascular Biology, 20*: 1990–1997; A. Minihane (2010). Fatty acid-genotype interactions and cardiovascular risk. *Prostaglandins, Leukotrienes and Essential Fatty Acids, 82*: 259–264.

7 M. Park (2011, April 13). Half of Americans use supplements. *CNN.com.*

8 H. Bastion (2008). Lucy Wills (1888–1964): The life and research of an adventurous independent woman. *The Journal of the Royal College of Physicians of Edinburgh, 38*: 89–91.

9 M. Hall (2012). *Mish-Mash of Marmite: A–Z of Tar-in-a-Jar.* London: BeWrite Books.

10 If you'd like to read more about these findings, see: P. Surén et al. (2013). Association between maternal use of folic acid supplements and risk of autism spectrum disorders in children. *The Journal of the American Medical Association, 309*: 570–577.

11 L. Yan et al. (2012). Association of the maternal *MTHFR* C677T polymorphism with susceptibility to neural tube defects in offsprings: Evidence from 25 case-control studies. *PLOS One, 7*: e41689.

12 A. Keller et al. (2012). New insights into the Tyrolean Iceman's origin and phenotype as inferred by whole-genome sequencing. *Nature Communications, 3*: 698.

13 I can't guarantee that signing up for the service won't result in a visit by LDS church missionaries: www.familysearch.org.

CHAPTER 7: PICKING SIDES

1 If you're not a surfing fan, you might remember Occhilupo from his stint on *Dancing with the Stars*. To learn more of the incredible story that came before he quickstepped his way into elimination on that popular TV show, read: M. Occhilupo and T. Baker (2008). *Occy: The Rise and Fall and Rise of Mark Occhilupo*. Melbourne: Random House Australia.

2 P. Hilts (1989, Aug. 29). A sinister bias: New studies cite perils for lefties. *The New York Times*.

3 L. Fritschi et al. (2007). Left-handedness and risk of breast cancer. *British Journal of Cancer, 5*: 686–687.

4 If you'd like to see the Walt Disney short *Hawaiian Holiday*, go to the following link: www.youtube.com/watch?v=SdIaEQCUVbk.

5 E. Domellöf et al. (2011). Handedness in preterm born children: A systematic review and a meta-analysis. *Neuropsychologia, 49*: 2299–2310.

6 If you're interested in learning more about this topic than you can read more: O. Basso (2007). Right or wrong? On the difficult relationship between epidemiologists and handedness. *Epidemiology,* 18: 191–193.

7 A. Rodriguez et al. (2010). Mixed-handedness is linked to mental health problems in children and adolescents. *Pediatrics, 125*: e340–e348.

8 G. Lynch et al. (2001). *Tom Blake: The Uncommon Journey of a Pioneer Waterman*. Irvine: Croul Family Foundation.

9 M. Ramsay (2010). Genetic and epigenetic insights into fetal alcohol spectrum disorders. *Genome Medicine, 2*: 27; K. R. Warren and T. K. Li. (2005).

Genetic polymorphisms: Impact on the risk of fetal alcohol spectrum disorders. *Birth Defects Research Part A: Clinical and Molecular Teratology, 73*: 195–203.

10 E. Domellöf et al. (2009). Atypical functional lateralization in children with fetal alcohol syndrome. *Developmental Psychobiology,* 51: 696–705.

11 Naranjo's story is nothing short of amazing. Be sure to check out the videos of him at work on YouTube, and don't miss: B. Edelman (2002, July 2). Michael Naranjo: The artist who sees with his hands. *Veterans Advantage.* http://www.veteransadvantage.com/cms/content/michael-naranjo

12 S. Moalem et al. (2013). Broadening the ciliopathy spectrum: Motile cilia dyskinesia, and nephronophthisis associated with a previously unreported homozygous mutation in the *INVS/NPHP2* gene. *American Journal of Medical Genetics Part A, 161*:1792–1796.

13 Did the meteorite simply pick up an extra smattering of aminos when it hit the lake? The scientists accounted for that: D. P. Glavin et al. (2012). Unusual nonterrestrial l-proteinogenic amino acid excesses in the Tagish Lake meteorite. *Meteoritics & Planetary Science, 47*: 1347–1364.

14 S. N. Han et al. (2004). Vitamin E and gene expression in immune cells. *Annals of the New York Academy of Sciences,1031*: 96–101.

15 G. J. Handleman et al. (1985). Oral alpha-tocopherol supplements decrease plasma gamma-tocopherol levels in humans. *The Journal of Nutrition, 115*: 807–813.

16 J. M. Major et al. (2012). Genome-wide association study identifies three common variants associated with serologic response to vitamin E supplementation in men. *The Journal of Nutrition, 142*: 866–871.

CHAPTER 8: WE'RE ALL X-MEN

1 For more information, visit the National Geographic Project: www. nationalgeographic.com

2 M. Hanaoka et al. (2012). Genetic variants in *EPAS1* contribute to adaptation to high-altitude hypoxia in Sherpas. *PLOS One, 7*: e50566.

3 One of the signs that pilots and aircrew members watch out for is an unexpected fit of giggling, which can be a sign that less oxygen is available due to an aircraft's fuselage becoming depressurized.

4 P. H. Hackett (2010). Caffeine at high altitude: Java at base camp. *High Altitude Medicine & Biology, 11*: 13–17.

5 Coca-Cola's slogan back in the mid-1940s.

6 A. de La Chapelle et al. (1993). Truncated erythropoietin receptor causes dominantly inherited benign human erythrocytosis. *Proceedings of the National Academy of Sciences, 90*: 4495–4499.

7 Apa Sherpa has made yearly returns to Nepal several times to raise awareness of climate change and the desperate need for better education in the Sherpa community, since moving to the United States with his wife and children in 2006. To read more about Apa Sherpa see the following article: M. LaPlante (2008, June 2). Everest record-holder proudly calls Utah home. *The Salt Lake Tribune*.

8 D. J. Gaskin et al. (2012). The economic costs of pain in the United States. *The Journal of Pain, 13*: 715–724.

9 B. Huppert (2011, Feb. 9). Minn. girl who feels no pain, Gabby Gingras, is happy to "feel normal." KARE11; K. Oppenheim (2006, Feb. 3). Life full of danger for little girl who can't feel pain. *CNN.com*.

10 J. J. Cox et al. (2006). An *SCN9A* channelopathy causes congenital inability to experience pain. *Nature*, 444: 894–898.

CHAPTER 9: HACKING YOUR GENOME

1 If you'd like more information regarding statistics related to the prevalence of many different types of cancer, the American Cancer Society website is a good place to start: www.cancer.org.

2 C. Brown (2009, Apr.). The king herself. *National Geographic, 215*(4).

3 It is still unclear what exact role diet played in the development of cancer in certain species of dinosaurs, since not all species seemed to be equally affected. If you'd like to read more about this fascinating work, see: B. M. Rothschild et al. (2003). Epidemiologic study of tumors in dinosaurs. *Naturwissenschaften, 90*: 495–500, and J. Whitfield (2003, Oct. 21). Bone scans reveal tumors only in duck-billed species. *Nature News*.

4 World Health Organization.

5 For more information about the rates and causes of lung cancer, see the Centers for Disease Control and Prevention's website, www.cdc.gov.

6 A. Marx. (1994–1995, Winter). The ultimate cigar aficionado. *Cigar Aficionado*.

7 This in spite of the fact that many of these publications were well funded by cigarette advertising.

8 R. Norr. (1952, December). Cancer by the carton. *The Reader's Digest.*

9 If you're interested in additional historical figures related to smoking, see the website www.lung.org.

10 *See It Now*, (1955, June 7). Transcribed from a tape recording made for Hill and Knowlton, Inc., during the telecast on CBS-TV.

11 U.S. Department of Agriculture. (2007). Tobacco Situation and Outlook Report Yearbook; Centers for Disease Control and Prevention. *National Center for Health Statistics. National Health Interview Survey 1965–2009.*

12 The entire transcript of "Cigarettes and Lung Cancer" from the June 7, 1955, edition of *See It Now* can be found online at the Legacy Tobacco Documents Library's website, www.legacy.library.ucsf.edu/tid/ppq36b00.

13 There's a lot of speculation about what saber-toothed cats (they weren't actually tigers) hunted, but researchers have noted that they were in the right place at the right time to chow down on some of our earliest ancestors: L. de Bonis et al. (2010). New saber-toothed cats in the Late Miocene of Toros Menalla (Chad). *Comptes Rendus Palevol., 9*: 221–227.

14 B. Ramazzini (2001). *De Morbis Artificum Diatriba. American Journal of Public Health, 91*: 1380–1382.

15 T. Lewin (2001, February 10). Commission sues railroad to end genetic testing in work injury cases. *The New York Times.*

16 P. A. Schulte and G. Lomax (2003). Assessment of the scientific basis for genetic testing of railroad workers with carpal tunnel syndrome. *Journal of Occupational and Environmental Medicine, 45*: 592–600.

17 These were generally families with uncommon conditions, and the rarity of the disorders they harbored might have might it easier for researchers to identify them, but the ease with which the researchers were able to identify patients is nonetheless disconcerting: M. Gymrek et al. (2013). Identifying personal genomes by surname inference. *Science, 339*: 321–324.

18 J. Smith (2013, Apr. 16). How social media can help (or hurt) you in your job search. *Forbes.com.*

19 In the U.S., employers and health insurance providers are limited in the genetic information they can seek.

20 In 2012, however, the Presidential Commission for the Study of Bioethical Issues issued a report calling for such tests to be made illegal, citing widespread privacy concerns: S. Begley (2012, Oct. 11). Citing privacy concerns, U.S. panel urges end to secret DNA testing. *Reuters.*

21 A. Jolie (2013, May 14). My medical choice. *The New York Times.*

22 D. Grady et al. (2013, May 14). Jolie's disclosure of preventive mastectomy highlights dilemma. *The New York Times*.

CHAPTER 10: MAIL-ORDER CHILD

1 Wrecksite is the world's largest online database of shipwrecks, with information on the final resting places of more than 140,000 ships. It's also a treasure trove of information about what many of those ships were doing when they met their fateful ends: http://www.wrecksite.eu.

2 See: I. Donald (1974). Apologia: How and why medical sonar developed. *Annals of the Royal College of Surgeons of England, 54*: 132–140.

3 This story and many more about German submarines can be found at www.uboat.net.

4 R. Brooks. (2013, Mar. 4). China's biggest problem? Too many men. *CNN.com*.

5 Y. Chen et al. (2013). Prenatal sex selection and missing girls in China: Evidence from the diffusion of diagnostic ultrasound. *The Journal of Human Resources, 48*: 36–70.

6 At one point in American history—and not so long ago—clothing "experts" advised parents to dress boys in pink and girls in blue. But by the 1950s and '60s, the gender paradigm had flipped. It might have flipped back or changed completely just as in-fashion colors do for adults had it not been for the advent of ultrasounds and sonograms: J. Paoletti (2012). *Pink and Blue: Telling the Boys from the Girls in America*. Indiana University Press.

7 This case presents a composite of previously published case reports and other similar patient encounters, with names, descriptions, and scenarios altered.

8 The Mayo Clinic's Disease Index has a detailed series of pages dedicated to hypospadias and thousands of other conditions: http://www.mayoclinic.com/health/DiseasesIndex.

9 It's possible that this is one of the most frequent autosomal recessive genetic disorders in human beings. P. W. Speiser et al. (1985). High frequency of nonclassical steroid 21-hydroxylase deficiency. *American Journal of Human Genetics, 37*: 650–667.

10 Like a clock, one arm is short (which we designate as "p") and the other is usually longer (we designate it as "q"). Each chromosome has a unique banding

pattern, which creates its barcode-like appearance under the microscope. It's these unique banding patterns that cytogeneticists use to identify and evaluate the integrity and quality of our chromosomes.

11 Unlike a karyotype, one of the important limitations of an aCGH is that it doesn't let you know if there's been a balanced movement or inversion of genetic material from one area in the genome to another. This is important because if we use the same example of encyclopedic volumes such a change can result in an entry being out of order, which for our genomes can be problematic. An aCGH cannot tell you if that has occurred.

12 Among other superstitions about *hijras*, many Indians believe that they must be present or nearby on wedding days to bring about good luck: N. Harvey (2008, May 13). India's transgendered—the Hijras. *New Statesman.*

13 Moreschi's complete recordings, which are scratchy and sometimes uneven but nonetheless spellbinding, are available on an 18-track compact disc, *The Last Castrato.* (1993). Opal.

14 K. J. Min et al. (2012). The lifespan of Korean eunuchs. *Current Biology, 22*: R792–R793.

15 Often wrongly attributed to Ralph Waldo Emerson, the slogan appears to have first appeared in a book by an anonymous securities trader whose identity was only years later publicly revealed by the *New York Times.* See H. Haskins (1940). *Meditations in Wall Street.* New York: William Morrow.

CHAPTER 11: PUTTING IT ALL TOGETHER

1 More than the entire population of the state of Texas: National Organization for Rare Disorders.

2 Fat gets a bad name. For most people, it's vital, and as this study found, the relationship between fat intake and reports of depression may be more complicated than we initially anticipated and may depend on the particular type of fat: A. Sánchez-Villegas et al. (2011). Dietary fat intake and the risk of depression: The SUN Project. *PLOS One, 26*: e16268.

3 Heart disease is sometimes called a "hidden" epidemic: D. L. Hoyert and J. Q. Xu. (2012). Deaths: Preliminary data for 2011. *National Vital Statistics Reports, 61*: 1–52.

4 S. C. Nagamani et al. (2012). Nitric-oxide supplementation for treatment of long-term complications in argininosuccinic aciduria. *American Journal of Human Genetics, 90*: 836–846; C. Ficicioglu et al. (2009). Argininosucci-

nate lyase deficiency: Longterm outcome of 13 patients detected by newborn screening. *Molecular Genetics and Metabolism, 98*: 273–277.

5 A. Williams (2013, Apr. 3). The Ecuadorian dwarf community "immune to cancer and diabetes" who could hold cure to diseases. *The Daily Mail.*

6 Gorlin syndrome isn't the only reason for toes webbed in this way. If you have syndactyly, it doesn't automatically mean you're likely to get skin cancer.

7 N. Boutet et al. (2003). Spectrum of *PTCH1* mutations in French patients with Gorlin syndrome. *The Journal of Investigative Dermatology, 121*: 478–481.

8 A. Case and C. Paxson (2006). *Stature and Status: Height, Ability, and Labor Market Outcomes.* National Bureau of Economic Research Working Paper No. 12466.

9 The French have long fought a losing battle against the idea that Napoléon was short and that his height played a factor in his empiric ambitions: M. Dunan (1963). La taille de Napoléon. *La Revue de l'Institut Napoléon, 89:* 178–179.

10 V. Ayyar (2011). History of growth hormone therapy. *Indian Journal of Endocrinology and Metabolism, 15*: S162–S165.

11 A. Rosenbloom (2011). Pediatric endo-cosmetology and the evolution of growth diagnosis and treatment. *The Journal of Pediatrics, 158*: 187–193.

INDEX

ACKNOWLEDGMENTS

I am grateful and indebted to all the patients and their families who have let me retell the stories of their medical journeys throughout the pages of *Inheritance*. I'm also extremely grateful to all the teachers and mentors I've had over the years, in medicine and beyond. I'm especially thankful to David Chitayat, MD, whose continual and inspirational support and enthusiasm for this project from its very early infancy was so crucial to its final success; over the years he has also generously shared with me his contagious passion for dysmorphology, genetics, and medicine. My agent Richard Abate of 3 Arts believed in this project from the start and was instrumental in conveying the importance of capturing, "how a geneticist thinks." The manuscript was immensely improved by the suggestions and direction of many readers. I must especially acknowledge my wonderful executive editor Ben Greenberg at Grand Central Publishing, whose intellectual probing and persistence helped bring clarity to complex genetic processes and ideas. Ben was also one of the earliest champions of *Inheritance* and was critical in ensuring the book received the audience he believed it deserves. I also would like to thank Drummond Moir, my UK editor at Sceptre for some last minute editorial pinch-hitting and helpful suggestions. To Yasmin Mathew, for her meticulous work as the production editor. And to Melissa Khan of 3 Arts, and Pippa White at Grand Central for always staying one step ahead administratively, and making it a surprising pleasure to meet deadlines. As well, to my publicists Matthew Ballast at Grand Central, as well as Catherine Whiteside, who both did such a wonderful job in raising essential awareness about the book. My research assistant Richard Verver continues to astound me with his unwaveringly sharp eye, and his relentless pursuit of original sources regardless of any language barriers. To Alaina deHavillard of Wailele Estates Kona Coffee whose masterful brew inspired page after page of this book. And to Wally whose gracious hospitality and welcoming home created the perfect ambiance to finish this project. As well, a special thank you to Jordan Peterson who spent an immense amount of time and energy on suggestions regarding the refinement of the manuscript. And of course to Matthew LaPlante who elevated this whole project with his immense journalistic talent and his refreshing sense of humor. Last but never least to my family and friends, for your endless love, support, and constant enthusiasm for every new project and undertaking.

Join a literary community of
like-minded readers who seek out
the best in contemporary writing.

From the thousands of submissions Sceptre
receives each year, our editors select the books
we consider to be outstanding.

We look for distinctive voices, thought-provoking
themes, original ideas, absorbing narratives and
writing of prize-winning quality.

If you want to be the first to hear about our
new discoveries, and would like the chance to
receive advance reading copies of our books
before they are published, visit

www.sceptrebooks.co.uk

Follow @sceptrebooks

'Like' SceptreBooks

Watch SceptreBooks